D0410654

6/74

MATRICES FOR SCIENTISTS

Applied Mathematics

Editor

I. W. ROXBURGH

Professor of Mathematics, Queen Mary College
University of London

MATRICES FOR SCIENTISTS

I. P. Williams
Reader in Mathematics, Queen Mary College
University of London

HUTCHINSON UNIVERSITY LIBRARY
LONDON

HUTCHINSON & CO *(Publishers)* LTD
3 Fitzroy Square, London W1

London Melbourne Sydney Auckland
Wellington Johannesburg Cape Town
and agencies throughout the world

First published 1972

This book has been set in cold type by E. W. C. Wilkins and Associates Ltd., printed in Great Britain by Anchor Press, and bound by Wm. Brendon, both of Tiptree, Essex

ISBN 0 09 110250 2 (cased)
0 09 110251 0 (paper)

CONTENTS

Preface 9

1. THE ALGEBRA OF MATRICES 11

1.1. Definition and notation 11
1.2. Multiplication 14
1.3. The determinant of a matrix 17
1.4. The transpose of a matrix 18
1.5. The transpose of the product of matrices 18
1.6. Addition of matrices 20
1.7. Special matrices 21
1.8. Useful relationships involving special matrices 27
Exercises 1 36

2. THE INVERSE OF A MATRIX 38

2.1. The existence of an inverse matrix 38
2.2. Minors and cofactors 39
2.3. The adjoint matrix 40
2.4. The inverse matrix 40
2.5. The inverse of the transpose of a matrix 43
2.6. The inverse of the product of matrices 43
2.7. The Gauss – Jordan method for calculating the inverse matrix 44
2.8. Pivoting 46
2.9. Improving an approximate inverse 48
2.10. Partitioned matrix 50
2.11. Calculating the inverse by means of partitioning 52

2.12. Application of partitioning to linear equations 54
Exercises 2 55

3. LINEAR EQUATIONS 57

3.1. Introduction 57
3.2. Homogeneous equation 57
3.3. Linear dependence 59
3.4. Solution of homogeneous equations and rank 61
3.5. Non-homogeneous equations 62
3.6. Gaussian elimination 63
3.7. Back substitution 65
3.8. Pivoting 65
3.9. Decomposition or **L U** method 67
3.10. Iterative methods for obtaining a solution 70
3.11. The Gauss – Seidel method 72
Exercises 3 74

4. EIGENVALUES AND EIGENVECTORS 77

4.1. The eigenvalues of a matrix 77
4.2. Eigenvectors 78
4.3. Reduction to a diagonal matrix 81
4.4. Linear dependence of the eigenvectors 81
4.5. Orthogonal reduction to a diagonal matrix 86
4.6. Geometric application 92
4.7. Spectral resolution 96
4.8. Functions of a matrix 97
4.9. Exponential of a matrix 98
Exercises 4 100

5. NUMERICAL DETERMINATION OF EIGENVALUES AND EIGENVECTORS 102

5.1. Introduction 102
5.2. The arbitrary vector method for determining the largest eigenvalue 102
5.3. The breakdown of the above method 105
5.4. The Rayleigh – Schwartz method for a symmetric matrix 106
5.5. The breakdown of the Rayleigh – Schwartz method 110
5.6. A useful elimination procedure 111

5.7. Evaluating the smaller eigenvalues 115
5.8. Jacobi methods 116
Exercises 5 119

6. SOME MATRIX APPLICATIONS 121

6.1. Introduction 121
6.2. Applications of matrix algebra 121
6.3. Use of eigenvalue properties 130

Further reading 142

Index 143

PREFACE

With the advent of the electronic computer, Matrix Theory is a topic that has found its way into most scientific fields. This book originated from notes of lectures given to practising scientists from many disciplines on several occasions, and also from undergraduate teaching. It is intended to give a reasonable coverage of matrix theory while at the same time bearing in mind that the matrices are to be used on a computer. The numerical methods that have been included are all fairly efficient and easy to follow and programmes for most of these will be found in any computer library, but they are not necessarily the most efficient or sophisticated method of tackling some problems.

I should like to thank Mrs L. Parry for her care and patience in typing the manuscript.

I. P.W.

I

THE ALGEBRA OF MATRICES

1.1. Definition and notation

In many branches of science and engineering, sets of linear equations of the general form

$$
\begin{aligned}
a_1 x + b_1 y + c_1 z &= x' \\
a_2 x + b_2 y + c_2 z &= y' \qquad (1.1) \\
a_3 x + b_3 y + c_3 z &= z'
\end{aligned}
$$

frequently occur. In these equations a_1, a_2, a_3, b_1, b_2, b_3, c_1, c_2 and c_3 are all assumed to be known constants, while x, y and z are the unknowns which are to be found provided that x', y' and z' are given. Alternatively, the set of equations may represent a transformation from the variables x, y, z to the variables x', y', z'. Examples of such sets of equations taken from genuine scientific problems will be found in Chapter 6.

It is clear that more economical ways of writing these equations must exist, for in the set of equations (1.1) x, y and z have all been written three times, while three equality signs and six addition signs have also been used. This repetition may not appear to be particularly significant with a set of only three equations, but consider the corresponding repetition that occurs in a set of say 120 equations with 120 unknowns. It is not uncommon to have to consider such a large number of equations in certain types of engineering and scientific problems.

In vector algebra it is customary to denote the scalar product of the two vectors $(a_1 b_1 c_1)$ and $(x\ y\ z)$ by

$$(a_1\ b_1\ c_1) \cdot \begin{pmatrix} x \\ y \\ z \end{pmatrix} = a_1 x + b_1 y + c_1 z \tag{1.2}$$

Adopting a similar convention here reduces the first equation of set (1.1) to

$$(a_1\, b_1\, c_1) \begin{pmatrix} x \\ y \\ z \end{pmatrix} = x'$$

A similar reduction is possible for the remaining two equations and the whole set of equations could be written as

$$\begin{pmatrix} a_1 & b_1 & c_1 \\ a_2 & b_2 & c_2 \\ a_3 & b_3 & c_3 \end{pmatrix} \begin{pmatrix} x \\ y \\ z \end{pmatrix} = \begin{pmatrix} x' \\ y' \\ z' \end{pmatrix} \tag{1.3}$$

provided that we interpret the left-hand side of this equation as meaning that each row of

$$\begin{pmatrix} a_1 & b_1 & c_1 \\ a_2 & b_2 & c_2 \\ a_3 & b_3 & c_3 \end{pmatrix}$$

is regarded as a vector and that the scalar product of each is taken in turn with the vector $\begin{pmatrix} x \\ y \\ z \end{pmatrix}$

With a set of equations of the type (1.1), corresponding to every set of values x', y' and z', there exist solutions which are values of x, y and z. Let the solution corresponding to x_1', y_1' and z_1' be x_1, y_1 and z_1 while the solution corresponding to x_2', y_2' and z_2' is x_2, y_2 and z_2. Then for the first set of values

$$\begin{pmatrix} a_1 & b_1 & c_1 \\ a_2 & b_2 & c_2 \\ a_3 & b_3 & c_3 \end{pmatrix} \begin{pmatrix} x_1 \\ y_1 \\ z_1 \end{pmatrix} = \begin{pmatrix} x_1' \\ y_1' \\ z_1' \end{pmatrix} \tag{1.4}$$

while for the second set of values

$$\begin{pmatrix} a_1 & b_1 & c_1 \\ a_2 & b_2 & c_2 \\ a_3 & b_3 & c_3 \end{pmatrix} \begin{pmatrix} x_2 \\ y_2 \\ z_2 \end{pmatrix} = \begin{pmatrix} x_2' \\ y_2' \\ z_2' \end{pmatrix} \qquad (1.5)$$

It would save considerable space if equations (1.4) and (1.5) could be combined and clearly this may be done as follows:

$$\begin{pmatrix} a_1 & b_1 & c_1 \\ a_2 & b_2 & c_2 \\ a_3 & b_3 & c_3 \end{pmatrix} \begin{pmatrix} x_1 & x_2 \\ y_1 & y_2 \\ z_1 & z_2 \end{pmatrix} = \begin{pmatrix} x_1' & x_2' \\ y_1' & y_2' \\ z_1' & z_2' \end{pmatrix} \qquad (1.6)$$

provided the convention is adopted that the first column in the second array on the left-hand side of the equation, that is

$\begin{pmatrix} x_1 \\ y_1 \\ z_1 \end{pmatrix}$, corresponds to the first column on the right-hand side of

the equation, that is $\begin{pmatrix} x_1' \\ y_1' \\ z_1' \end{pmatrix}$, and similarly for the second columns.

The three arrays in equation (1.6) namely

$$\begin{pmatrix} a_1 & b_1 & c_1 \\ a_2 & b_2 & c_2 \\ a_3 & b_3 & c_3 \end{pmatrix}, \begin{pmatrix} x_1 & x_2 \\ y_1 & y_2 \\ z_1 & z_2 \end{pmatrix} \text{ and } \begin{pmatrix} x_1' & x_1' \\ y_1' & y_2' \\ z_1' & z_2' \end{pmatrix}$$

are each examples of a MATRIX. The first is said to be a 3×3 matrix as it has 3 rows and 3 columns. It may also be referred to as a square matrix of order 3. The second and third arrays are both 3×2 matrices as they have 3 rows and 2 columns. A matrix may therefore be DEFINED to be any ordered array of numbers such as the three examples given above. The actual numbers in the ordered array are usually called the ELEMENTS of the matrix.

Two properties of a matrix are now implicitly contained in equation (1.6). First, it defines the rule for the multiplication of two matrices. In order to obtain the element in row i and column j of the product, row i of the left-hand matrix is regarded as one vector while column j of the right-hand matrix is regarded as another vector and their scalar product is calculated. Repetition for every possible value of i and j determines all of the elements of a matrix which is the product of the two original matrices. The second property that is implied in equation (1.6) is that concerning the use of the equality sign. For equation (1.6) to be consistent, two matrices can be equal only if all the corresponding elements are equal. In future we shall assume that the equality sign implies exactly this. One useful, but trivial, corollary of the above definition of equality is that two matrices can be equal if and only if they have the same number of rows and the same number of columns.

While the notation so far employed is adequate for small matrices, it is clearly inadequate for a matrix with more than 26 columns as no more letters will be available to denote further columns. The simplest way to overcome this difficulty is to use two subscripts; the first subscript denotes which row the element is in while the second subscript denotes the column. Using this notation equation (1.6) can be rewritten as

$$\begin{pmatrix} a_{11} & a_{12} & a_{13} \\ a_{21} & a_{22} & a_{23} \\ a_{31} & a_{32} & a_{33} \end{pmatrix} \begin{pmatrix} x_{11} & x_{12} \\ x_{21} & x_{22} \\ x_{31} & x_{32} \end{pmatrix} = \begin{pmatrix} x'_{11} & x'_{12} \\ x'_{21} & x'_{22} \\ x'_{31} & x'_{32} \end{pmatrix} \qquad (1.7)$$

To reduce writing still further, a single capital letter in heavy type can be used to denote a matrix, **A** say, its general element then being the corresponding lower case letter with two subscripts, a_{ij}. Equation (1.7) can therefore be written as

$$\mathbf{A}\ \mathbf{X} = \mathbf{X}'$$

where **A** has the general element a_{ij}, **X** the general element x_{ij} and **X**$'$ the general element x'_{ij}.

1.2. Multiplication

As has already been discussed, the rule for the multiplication of two matrices was implicit in equation (1.6). Using the notation that has just been defined, if we have three matrices **A**, **B** and **C** with their respective general elements as a_{ij}, b_{ij} and c_{ij}, then

$$\mathbf{A}\ \mathbf{B} = \mathbf{C}$$

implies

$$\sum_k a_{ik} b_{kj} = c_{ij} \tag{1.8}$$

which states that the element in row i and column j of **C** is equal to the scalar product of row i of **A** and column j *of* **B**.

From the definition of multiplication as expressed in equation (1.8) the following general rules can be deduced.

(a) For the product **AB** to exist, the number of columns in **A** has to be equal to the number of rows in **B**.

(b) If **A B** = **C**, then the number of rows in **A** is the same as the number of rows in **C** while the number of columns in **B** is the same as the number of columns in **C**.

(c) The product **A B** need not equal the product **B A**. Care must therefore be taken to maintain the order of the matrices in any multiplication process.

(d) If there exists a scalar factor λ that is common to all the elements of **A** then it may be taken outside the matrix, so that

$$\begin{pmatrix} \lambda a_{11} & \lambda a_{12} & \cdots & \lambda a_{1n} \\ \cdots & \cdots & \cdots & \cdots \\ \lambda a_{m1} & \lambda a_{m2} & \cdots & \lambda a_{mn} \end{pmatrix} = \lambda \begin{pmatrix} a_{11} & a_{12} & \cdots & a_{1n} \\ \cdots & \cdots & \cdots & \cdots \\ a_{m1} & a_{m2} & \cdots & a_{mn} \end{pmatrix} \tag{1.9}$$

(e) If all the above rules for multiplication are satisfied, then

$$\mathbf{A}\ (\mathbf{B}\ \mathbf{C}) = (\mathbf{A}\ \mathbf{B})\ \mathbf{C}$$

The proofs of the first four statements are all very obvious and involve only writing down definition (1.8) in a suitable form. For this reason they have not been included. Statement (e) is slightly more complex and is now proved. Let **A** (**B C**) = **D** while (**A B**) **C** = **E**. In order to prove the statement it is therefore required to show that **D** = **E**.

$$d_{ij} = \sum_s a_{is} \sum_k b_{sk} c_{kj}$$

on using the defined multiplication rule twice. But a_{is} is independent of k and so

$$d_{ij} = \sum_s \sum_k a_{is}\, b_{sk}\, c_{kj}$$

By using the multiplication rule as before we have

$$e_{ij} = \sum_k \left(\sum_s a_{is}\, b_{sk}\right) c_{kj} = \sum_k \sum_s a_{is}\, b_{sk}\, c_{kj}$$

However all the elements are numbers and so

$$e_{ij} = \sum_s \sum_k a_{is}\, b_{sk}\, c_{kj} = d_{ij}$$

Thus all the elements of **E** are equal to the corresponding elements of **D** and the statement has been proved.

Example 1.1: Calculate the product **A B**, if it exists, given that

$$\mathbf{A} = \begin{pmatrix} 1 & 2 & 3 \\ 4 & 5 & 6 \\ 7 & 8 & 9 \end{pmatrix} \quad \text{and} \quad \mathbf{B} = \begin{pmatrix} 1 & 1 & 1 \\ 1 & 0 & 1 \\ 0 & 1 & 1 \end{pmatrix}$$

$$\mathbf{A\,B} = \begin{pmatrix} 1 & 2 & 3 \\ 4 & 5 & 6 \\ 7 & 8 & 9 \end{pmatrix} \begin{pmatrix} 1 & 1 & 1 \\ 1 & 0 & 1 \\ 0 & 1 & 1 \end{pmatrix} =$$

$$\begin{pmatrix} 1.1 + 2.1 + 3.0 & 1.1 + 2.0 + 3.1 & 1.1 + 2.1 + 3.1 \\ 4.1 + 5.1 + 6.0 & 4.1 + 5.0 + 6.1 & 4.1 + 5.1 + 6.1 \\ 7.1 + 8.1 + 9.0 & 7.1 + 8.0 + 9.1 & 7.1 + 8.1 + 9.1 \end{pmatrix}$$

$$\begin{pmatrix} 3 & 4 & 6 \\ 9 & 10 & 15 \\ 15 & 16 & 24 \end{pmatrix}$$

Example 1.2: Calculate the product **A B** and the product **B A**, if they exist, when

$$\mathbf{A} = \begin{pmatrix} 1 & 0 & 1 \\ 0 & 1 & 0 \\ 1 & 0 & 3 \end{pmatrix} \quad \text{and} \quad \mathbf{B} = \begin{pmatrix} 1 & 4 \\ 2 & 5 \\ 3 & 6 \end{pmatrix}$$

$$\mathbf{A\,B} = \begin{pmatrix} 1 & 0 & 1 \\ 0 & 1 & 0 \\ 1 & 0 & 3 \end{pmatrix} \begin{pmatrix} 1 & 4 \\ 2 & 5 \\ 3 & 6 \end{pmatrix} =$$

$$\begin{pmatrix} 1.1+0.2+1.3 & 1.4+0.5+1.6 \\ 0.1+1.2+0.3 & 0.4+1.5+0.6 \\ 1.1+0.2+3.3 & 1.4+0.3+3.6 \end{pmatrix} = \begin{pmatrix} 4 & 10 \\ 2 & 5 \\ 10 & 22 \end{pmatrix}$$

Since the number of columns in **B** is not equal to the number of rows in **A** the product **B A** cannot exist. This verifies the rule that **B A** need not equal **A B**.

1.3. The determinant of a matrix

The DETERMINANT of a matrix is defined to be the determinant whose elements are identical to those of the matrix. Clearly, as determinants are always square, the determinant of a matric can only exist if the matrix is square. We shall denote the determinant of a square matrix **A** by $|\mathbf{A}|$.

Since the rule for the multiplication of two determinants is identical to that which has been defined for the multiplication of two matrices, it follows that the product of the determinants of the matrices is equal to the determinant of the product of the matrices, that is

$$|\mathbf{A}|.|\mathbf{B}| = |\mathbf{A}\ \mathbf{B}| \tag{1.10}$$

The square matrix **A** is said to be SINGULAR if $|\mathbf{A}| = 0$ and otherwise is said to be NON-SINGULAR.

Example 1.3: Verify that $|\mathbf{A}|\ |\mathbf{B}| = |\mathbf{A}\ \mathbf{B}|$ for the matrices **A** and **B** given by

$$\mathbf{A} = \begin{pmatrix} 1 & 2 & 3 \\ 4 & 5 & 6 \\ 0 & 8 & 7 \end{pmatrix} \quad \text{and } \mathbf{B} = \begin{pmatrix} 0 & 0 & 1 \\ 0 & 1 & 0 \\ 1 & 0 & 0 \end{pmatrix}$$

$$\mathbf{A}\ \mathbf{B} = \begin{pmatrix} 1 & 2 & 3 \\ 4 & 5 & 6 \\ 0 & 8 & 7 \end{pmatrix} \begin{pmatrix} 0 & 0 & 1 \\ 0 & 1 & 0 \\ 1 & 0 & 0 \end{pmatrix} = \begin{pmatrix} 3 & 2 & 1 \\ 6 & 5 & 4 \\ 7 & 8 & 0 \end{pmatrix}$$

$$|\mathbf{A}\ \mathbf{B}| = \begin{vmatrix} 3 & 2 & 1 \\ 6 & 5 & 4 \\ 7 & 8 & 0 \end{vmatrix} = -27$$

$$|\mathbf{A}| = \begin{vmatrix} 1 & 2 & 3 \\ 4 & 5 & 6 \\ 0 & 8 & 7 \end{vmatrix} = 27 \text{ and } |\mathbf{B}| = \begin{vmatrix} 0 & 0 & 1 \\ 0 & 1 & 0 \\ 1 & 0 & 0 \end{vmatrix} = -1$$

Hence for these matrices $|\mathbf{A}| \,.\, |\mathbf{B}| = |\mathbf{A}\,\mathbf{B}| = -27$. Since $|\mathbf{A}| = 27 = 0$ and $|\mathbf{B}| = -1 \neq 0$, both **A** and **B** are examples of non-singular matrices.

1.4. The transpose of a matrix

The TRANSPOSE of a matrix is the matrix that is obtained when all rows of the original matrix are changed into columns and, automatically, all the columns into rows. We shall denote this operation by a superscript T so that, if **A** is the original matrix, $\mathbf{A}^T$ is the matrix whose rows are the columns of **A**. If a_{ij} is the element in row i and column j of the matrix **A**, then in row i and column j *of* $\mathbf{A}^T$ the element will be a_{ij}.

The transpose symbol is very useful as a space saver when a problem contains large column vectors because the vector

$$\begin{pmatrix} x_1 \\ x_2 \\ \vdots \\ x_n \end{pmatrix} = (x_1, x_2, \ldots, x_n^T)$$

Any column vector can therefore be written as the transpose of a row vector.

The use of the transpose operator is also of considerable use where multiplications are being carried out on an electronic computer. For technical reasons it is more efficient for the computer either to sweep along rows only or along columns only. If the product **AB** is required, then the matrices **A** and $\mathbf{B}^T$ are fed into the computer. Sweeping along the rows of **A** followed by the *rows* of $\mathbf{B}^T$ is then equivalent to taking the rows of **A** and the columns of **B**, the correct procedure in order to obtain the product.

1.5. The transpose of the product of matrices

If $\mathbf{A}\,\mathbf{B} = \mathbf{C}$ then the problem is to express $\mathbf{C}^T$ in terms of $\mathbf{A}^T$ and $\mathbf{B}^T$. The general element in row i and column j of $\mathbf{C}^T$ is the element in row j and column i of $\mathbf{C}$ and from the definition of multiplication this is given by

$$c_{ij}^T = c_{ji} = \sum_k a_{jk} \, b_{ki}$$

However, since both a_{jk} and b_{ki} represent sequences of pure numbers,

$$\sum_k a_{jk} \, b_{ki} = \sum_k b_{ki} \, a_{jk}$$

But

$$\sum_k b_{ki} \, a_{jk} = \sum_k b_{ik}^T \, a_{kj}^T$$

which is the general term originating from the multiplication $\mathbf{B}^T\mathbf{A}^T$. This proves that

$$(\mathbf{A} \; \mathbf{B})^T = \mathbf{C}^T = \mathbf{B}^T\mathbf{A}^T \tag{1.11}$$

thus giving the rule for obtaining the transpose of the product of matrices in terms of the product of the transpose of matrices. The extension to any number of matrices is obvious, giving for four matrices, for example,

$$(\mathbf{A} \; \mathbf{B} \; \mathbf{C} \; \mathbf{D})^T = (\mathbf{C} \; \mathbf{D})^T (\mathbf{A} \; \mathbf{B})^T$$

on using the above rule. Using it again on $(\mathbf{C} \; \mathbf{D})^T$ and $(\mathbf{A} \; \mathbf{B})^T$ gives

$$(\mathbf{A} \; \mathbf{B} \; \mathbf{C} \; \mathbf{D})^T = \mathbf{D}^T\mathbf{C}^T\mathbf{B}^T\mathbf{A}^T$$

Using the same rule a sufficient number of times gives the product of n matrices as

$$(\mathbf{A}_1 \; \mathbf{A}_2 \ldots \; \mathbf{A}_n)^T = \mathbf{A}_n^T \ldots \; \mathbf{A}_2^T \, \mathbf{A}_1^T$$

Example 1.4: Verify that $(\mathbf{A} \; \mathbf{B})^T = \mathbf{B}^T\mathbf{A}^T$ when

$$\mathbf{A} = \begin{pmatrix} 1 & 2 & 3 \\ 4 & 5 & 6 \\ 7 & 8 & 9 \end{pmatrix} \quad \text{and} \quad \mathbf{B} = \begin{pmatrix} 1 & 1 & 1 \\ 1 & 1 & 1 \\ 1 & 1 & 1 \end{pmatrix}$$

$$\mathbf{A} \; \mathbf{B} = \begin{pmatrix} 1 & 2 & 3 \\ 4 & 5 & 6 \\ 7 & 8 & 9 \end{pmatrix} \begin{pmatrix} 1 & 1 & 1 \\ 1 & 1 & 1 \\ 1 & 1 & 1 \end{pmatrix} = \begin{pmatrix} 6 & 6 & 6 \\ 15 & 15 & 15 \\ 24 & 24 & 24 \end{pmatrix}$$

Transposing we obtain

$$(\mathbf{A} \; \mathbf{B})^T = \begin{pmatrix} 6 & 6 & 6 \\ 15 & 15 & 15 \\ 24 & 24 & 24 \end{pmatrix}^T = \begin{pmatrix} 6 & 15 & 24 \\ 6 & 15 & 24 \\ 6 & 15 & 24 \end{pmatrix}$$

But $\mathbf{B}^T = \begin{pmatrix} 1 & 1 & 1 \\ 1 & 1 & 1 \\ 1 & 1 & 1 \end{pmatrix}$ and $\mathbf{A}^T = \begin{pmatrix} 1 & 4 & 7 \\ 2 & 5 & 8 \\ 3 & 6 & 9 \end{pmatrix}$ and so

$$\mathbf{B}^T\mathbf{A}^T = \begin{pmatrix} 1 & 1 & 1 \\ 1 & 1 & 1 \\ 1 & 1 & 1 \end{pmatrix} \begin{pmatrix} 1 & 4 & 7 \\ 2 & 5 & 8 \\ 3 & 6 & 9 \end{pmatrix} = \begin{pmatrix} 6 & 15 & 24 \\ 6 & 15 & 24 \\ 6 & 15 & 24 \end{pmatrix}$$

thus verifying that $(\mathbf{A}\,\mathbf{B})^T = \mathbf{B}^T\mathbf{A}^T$.

1.6. Addition of matrices

The sum of two matrices is defined to be the matrix whose elements are the sum of the corresponding elements in the initial matrices. In the adopted notation this means that

$$\mathbf{C} = \mathbf{A} + \mathbf{B}$$

implies
$$c_{ij} = a_{ij} + b_{ij} \qquad (1.12)$$

This definition is meaningful only if all three matrices, **A**, **B** and **C**, are of the same order and so addition is only defined for similar matrices. Note that if individual rows of the matrices are regarded as vectors, then the given definition is consistent with the usual rule for the addition of vectors.

From the definition of addition as embodied in equation (1.12) the following two rules of addition can easily be deducted:

(a) $\mathbf{A} + \mathbf{B} = \mathbf{B} + \mathbf{A}$ for any two matrices of the same order

(b) $\mathbf{A} + (\mathbf{B} + \mathbf{C}) = (\mathbf{A} + \mathbf{B}) + \mathbf{C}$ for any three matrices of the same order

By combining the rule for addition given above and that for multiplication given in section 1.2 it is also possible to demonstrate:

(c) $\mathbf{A}(\mathbf{B} + \mathbf{C}) = \mathbf{A}\,\mathbf{B} + \mathbf{A}\,\mathbf{C}$

Example 1.5: Evaluate $\mathbf{A} + \mathbf{B}$ when

$$\mathbf{A} = \begin{pmatrix} 1 & 2 \\ 3 & 4 \end{pmatrix} \quad \text{and} \quad \mathbf{B} = \begin{pmatrix} 4 & 3 \\ 2 & 1 \end{pmatrix}.$$

$$\mathbf{A}+\mathbf{B}=\begin{pmatrix}1 & 2\\ 3 & 4\end{pmatrix}+\begin{pmatrix}4 & 3\\ 2 & 1\end{pmatrix}=\begin{pmatrix}1+4 & 2+3\\ 3+2 & 4+1\end{pmatrix}=\begin{pmatrix}5 & 5\\ 5 & 5\end{pmatrix}=5\begin{pmatrix}1 & 1\\ 1 & 1\end{pmatrix}$$

The last step also illustrates rule (d) of section 1.2.

1.7. Special matrices

There are several types of matrices which, because of either their behaviour or their appearance, have been given distinctive names. The most important of these special matrices are discussed briefly below. Useful interrelationships between these matrices and other matrices will be discussed in section 1.8.

Null matrix

This, as its name suggests, is the matrix which corresponds to zero in normal arithmetic for both the addition and the multiplication of matrices. It has all its elements equal to zero, that is

$$a_{ij} = 0 \text{ for all } i \text{ and } j$$

We shall denote the null matrix by $\mathbf{O}$ and it is obvious that if $\mathbf{A}$ is any other matrix then

$$\mathbf{A}\,\mathbf{O} = \mathbf{O} \text{ and } \mathbf{A} + \mathbf{O} = \mathbf{A}.$$

It is sometimes also called the zero matrix.

Diagonal matrix

Any square matrix in which all the elements with the exception of those on the leading diagonal, that is the diagonal joining the top left-hand corner to the bottom right-hand corner, is called a diagonal matrix. Thus, if $\mathbf{A}$ is a diagonal matrix, then

$$a_{ij} = 0 \text{ for all } i \neq j$$

A typical example of a diagonal matrix of order 3 is

$$\begin{pmatrix}a_{11} & 0 & 0\\ 0 & a_{22} & 0\\ 0 & 0 & a_{33}\end{pmatrix}$$

A diagonal matrix is usually denoted by $\mathbf{diagA}$. In a certain context, which will be discussed in Chapters 4 and 5, it is also customary to use Λ to denote a diagonal matrix.

Unit matrix

A diagonal matrix in which all the diagonal elements are equal to unity is called the unit matrix and will be denoted by **I** in this book. It is easy to verify that this matrix plays the part which, for multiplication purposes, corresponds to unity in normal arithmetic, so that if **A** is any matrix,

$$\mathbf{A}\,\mathbf{I} = \mathbf{A} \text{ and } \mathbf{I}\,\mathbf{A} = \mathbf{A} \tag{1.13}$$

The unit matrix of order 3 is

$$\mathbf{I} = \begin{pmatrix} 1 & 0 & 0 \\ 0 & 1 & 0 \\ 0 & 0 & 1 \end{pmatrix}$$

Banded matrix

A banded matrix is any square matrix in which the only non-zero elements occur in a band about the diagonal. Thus, if **A** is to be a banded matrix,

$$a_{ij} = 0 \text{ when } |i\text{-}j| > k.$$

In this case only $2k+1$ elements in every row (but the first and last) are non-zero. A typical banded matrix of order 4 is

$$\mathbf{A} = \begin{pmatrix} a_{11} & a_{12} & 0 & 0 \\ a_{21} & a_{22} & a_{23} & 0 \\ 0 & a_{32} & a_{33} & a_{34} \\ 0 & 0 & a_{43} & a_{44} \end{pmatrix}$$

and its elements satisfy $a_{ij} = 0$ if $|i\text{-}j| > 1$. There is no universally accepted symbol to denote a banded matrix and so we will not adopt one here.

Inverse matrix (or the reciprocal matrix)

It is not possible to define a process of division for matrices. However, under certain circumstances it is possible to determine the matrix **B** which satisfies the equation

$$\mathbf{A}\,\mathbf{B} = \mathbf{C} \tag{1.14}$$

when **A** and **C** are given. This is in some sense equivalent to a division of **C** by **A**.

If some matrix, $\mathbf{A}^{-1}$ say, can be found such that

$$\mathbf{A}\,\mathbf{A}^{-1} = \mathbf{I} \text{ and } \mathbf{A}^{-1}\,\mathbf{A} = \mathbf{I} \tag{1.15}$$

then, on multiplying equation (1.14) on the left by this matrix $\mathbf{A}^{-1}$, we have

$$\mathbf{A}^{-1}\,\mathbf{A}\,\mathbf{B} = \mathbf{I}\,\mathbf{B} = \mathbf{B} = \mathbf{A}^{-1}\,\mathbf{C}$$

and the matrix $\mathbf{B}$ has been determined.

The matrix $\mathbf{A}^{-1}$ is called the inverse matrix and it need not always exist. A fuller discussion of this topic is given in Chapter 2.

Upper triangular matrix

This is a type of square matrix where all the non-zero elements occur either on the leading diagonal or in the triangular section above it. Such a matrix will usually be denoted by $\mathbf{U}$ and its general term must satisfy

$$u_{ij} = 0 \text{ for all } i > j$$

which specifies that all the terms below the leading diagonal are equal to zero.

An example of an upper triangular matrix of order 3 is

$$\mathbf{U} = \begin{pmatrix} u_{11} & u_{12} & u_{13} \\ 0 & u_{22} & u_{23} \\ 0 & 0 & u_{33} \end{pmatrix}$$

Unfortunately some authors have used $\mathbf{U}$ to denote the unit matrix. Throughout this book $\mathbf{U}$ will denote the upper triangular matrix and $\mathbf{I}$ the unit matrix.

Unit upper triangular matrix

As the name implies, this is a special case of an upper triangular matrix in which all the elements on the leading diagonal are equal to unity. Hence, in addition to

$$u_{ij} = 0 \text{ for all } i > j,$$

we now require

$$u_{ij} = 1 \text{ for all } i = j$$

A unit upper triangular matrix of order 3 therefore has the general appearance

$$\mathbf{U} = \begin{pmatrix} 1 & u_{12} & u_{13} \\ 0 & 1 & u_{23} \\ 0 & 0 & 1 \end{pmatrix}$$

Lower triangular matrix

In a similar way to the above, if all the non-zero elements occur either on the leading diagonal or in the triangular section below it, the matrix is called a lower triangular matrix and is usually denoted by **L**. Its general element must satisfy

$$l_{ij} = 0 \text{ for all } i < j$$

which states that all the elements above the diagonal are equal to zero. A typical lower triangular matrix of order 3 is

$$\mathbf{L} = \begin{pmatrix} l_{11} & 0 & 0 \\ l_{12} & l_{22} & 0 \\ l_{13} & l_{23} & l_{33} \end{pmatrix}$$

Unit lower triangular matrix

This is the special case of a lower triangular matrix in which all the diagonal elements are equal to unity. Its elements therefore satisfy

$$l_{ij} = 0 \text{ for all } i < j$$
$$l_{ij} = 1 \text{ for all } i = j$$

A unit lower triangular matrix of order 3 is

$$\mathbf{L} = \begin{pmatrix} 1 & 0 & 0 \\ l_{12} & 1 & 0 \\ l_{13} & l_{23} & 1 \end{pmatrix}$$

Symmetric matrix

A symmetric matrix is defined to be a matrix that remains invariant when transposed, that is

$$\mathbf{A}^T = \mathbf{A} \tag{1.16}$$

The general term of **A** must therefore satisfy

$$a_{ij} = a_{ji}$$

This implies that i and j must be able to take the same values and so the matrix must be square. It also indicates that the matrix is symmetrical about its leading diagonal, hence the terminology.

A typical example of such a matrix of order 3 is

$$\mathbf{A} = \begin{pmatrix} a_{11} & b & c \\ b & a_{22} & d \\ c & d & a_{33} \end{pmatrix}$$

Anti-symmetric matrix

By contrast to the above, an anti-symmetric matrix is defined to be a matrix that changes in sign when transposed, so that

$$\mathbf{A}^T = -\mathbf{A} \tag{1.17}$$

Again the matrix must be square and the general element now satisfies

$$a_{ij} = -a_{ji}$$

This equation shows that

$$a_{ij} = 0 \text{ for all } i = j$$

indicating that as the diagonal elements remain unaltered and so cannot change in sign, they must be zero.

An example of an anti-symmetric matrix of order 3 is

$$\mathbf{A} = \begin{pmatrix} 0 & a & b \\ -a & 0 & c \\ -b & -c & 0 \end{pmatrix}$$

This type of matrix is also sometimes called a skew-symmetric matrix.

Hermitian matrix

The hermitian matrix is a generalization of the symmetric matrix which is often more useful than a symmetric matrix if some of the elements are complex numbers. It is defined to be the matrix which satisfies

$$\mathbf{A}^T = \bar{\mathbf{A}}$$

where the bar denotes that the complex conjugate of each element of the matrix is taken. The general element of such a matrix must satisfy

$$a_{ji} = \bar{a}_{ij}$$

Since this implies that

$$a_{ii} = \bar{a}_{ii}$$

it shows that all the elements for which $i = j$, that is all the elements on the diagonal, must be real. A typical third order hermitian matrix therefore has the appearance

$$\begin{pmatrix} a_{11} & a+ib & c+id \\ a-ib & a_{22} & e+if \\ c-id & a-if & a_{33} \end{pmatrix}$$

where a_{ii}, a, b, c, d, e and f are all real.

Anti-hermitian matrix
In a similar way this is a generalization of an anti-symmetric matrix to the case when some of the elements are complex. An anti-hermitian matrix satisfies

$$\mathbf{A}^T = -\bar{\mathbf{A}}$$

and its general element satisfies

$$a_{ji} = -\bar{a}_{ij}$$

It therefore follows that

$$a_{ii} = -\bar{a}_{ii}$$

and so all the terms on the leading diagonal must be pure imaginary numbers.

An example of an anti-hermitian matrix of order 3 is

$$\mathbf{A} = \begin{pmatrix} ia_{11} & a+ib & c+id \\ -a+ib & ia_{22} & e+if \\ -c+id & -e+if & ia_{33} \end{pmatrix}$$

where a_{ii}, a, b, c, d, e and f are again all real. This matrix is also sometimes called a skew-hermitian matrix.

Orthogonal matrix
Before defining an orthogonal matrix, it is useful to recall two

terms that are used in the study of vector algebra which will be of considerable use here, namely NORM and ORTHOGONAL. There are a number of possible definitions for the NORM of a vector. We shall use the definition that it is magnitude of the vector and is therefore also given by the square root of the product of the vector and transpose.

Two vectors are said to be ORTHOGONAL if, and only if, their scalar product is zero, or, with the convention we have adopted, if the product of one vector and the transpose of the second is zero.

Now, since any row or any column in a matrix has all the characteristics of a vector, an orthogonal matrix is defined to be a square matrix satisfying the following two conditions:

(a) The norms of all its rows and all its columns are equal to unity.
(b) Any row is orthogonal to every other row in the matrix and similarly for the columns.

Combining these two conditions, it is clear that if $\mathbf{A}$ is an orthogonal matrix, then

$$\mathbf{A}^T \mathbf{A} = \mathbf{I} = \mathbf{A}^T \mathbf{A} \tag{1.18}$$

Hence an orthogonal matrix has the very useful property that its inverse matrix is the same as its transpose matrix, or

$$\mathbf{A}^{-1} = \mathbf{A}^T$$

1.8. Useful relationships involving special matrices

All the special matrices mentioned above are, in some respect, simpler than a general matrix. For some types of problems the amount of manipulative work involved may therefore be reduced if the general matrices could be replaced by a simple combination of special matrices. Below we list several possible combinations which have proved to be very useful in practice.

(a) *Any real square matrix can be expressed as the sum of a symmetric matrix and an anti-symmetric matrix*

Let $\mathbf{A}$ be the given matrix with its general element a_{ij}. It is required to express this in the form

$$\mathbf{A} = \mathbf{B} + \mathbf{C}$$

where **B** is a symmetric matrix with its general element b_{ij} and **C** is an anti-symmetric matrix with its general elements c_{ij}. The rules of addition of matrices gives

$$a_{ij} = b_{ij} + c_{ij}$$
$$\text{and} \quad a_{ji} = b_{ji} + c_{ji}$$

However, as **B** is a symmetric matrix, $b_{ij} = b_{ji}$, while $c_{ij} = -c_{ji}$ as **C** is an anti-symmetric matrix. The above two equations therefore become

$$a_{ij} = b_{ij} + c_{ij}$$
$$a_{ji} = b_{ij} - c_{ij}$$

As a_{ij} and a_{ji} are given it is therefore possible to calculate b_{ij} and c_{ij}, the elements of the symmetric and the anti-symmetric matrix, the solution to the above in fact being

$$b_{ij} = \frac{a_{ij} + a_{ji}}{2}$$

$$\text{and} \quad c_{ij} = \frac{a_{ij} - a_{ji}}{2} \tag{1.19}$$

Hence a symmetric matrix and an anti-symmetric matrix can be found such that their sum is equal to the given real square matrix.

Example 1.6:

Express the matrix $\begin{pmatrix} 1 & 5 & 6 \\ 3 & 2 & 10 \\ 2 & 4 & 3 \end{pmatrix}$ as the sum of a symmetric matrix

and an anti-symmetric matrix. Let the symmetric matrix be **A** and the anti-symmetric matrix be **B**, then let

$$\mathbf{A} = \begin{pmatrix} a_{11} & a_{12} & a_{13} \\ a_{11} & a_{22} & a_{23} \\ a_{13} & a_{23} & a_{33} \end{pmatrix} \text{and } \mathbf{B} = \begin{pmatrix} 0 & b_{12} & b_{13} \\ -b_{12} & 0 & b_{23} \\ -b_{13} & -b_{23} & 0 \end{pmatrix}$$

Thus

$$a_{12} = 1, \ a_{22} = 2 \text{ and } a_{33} = 3$$

Also,

$a_{12} + b_{12} = 5$, while $a_{12} - b_{12} = 3$, so $a_{12} = 4$ and $b_{12} = 1$

$a_{13} + b_{13} = 6$, while $a_{13} - b_{13} = 2$, so $a_{13} = 4$ and $b_{13} = 2$

$a_{23} + b_{23} = 10$, while $a_{23} - b_{23} = 4$, so $a_{23} = 7$ and $b_{23} = 3$

Substituting the elements into the matrices we therefore have

$$\begin{pmatrix} 1 & 5 & 6 \\ 3 & 2 & 10 \\ 2 & 4 & 3 \end{pmatrix} = \begin{pmatrix} 1 & 4 & 4 \\ 4 & 2 & 7 \\ 4 & 7 & 3 \end{pmatrix} + \begin{pmatrix} 0 & 1 & 2 \\ -1 & 0 & 3 \\ -2 & -3 & 0 \end{pmatrix}$$

giving the original matrix as the sum of a symmetric and an anti-symmetric matrix.

(b) *Any square matrix can be expressed as the sum of a hermitian matrix and an anti-hermitian matrix*

This is the generalization of statement (a) to the case when elements of the matrix may be complex. Let the given matrix have as general element $a_{ij} + ib_{ij}$, where a_{ij} and b_{ij} are both real. Similarly let the hermitian and anti-hermitian matrices have respectively the general elements $c_{ij} + id_{ij}$ and $e_{ij} + if_{ij}$, where c_{ij}, d_{ij}, e_{ij} and f_{ij} are all real. Using the given rule for the addition of matrices and equating real parts, we have

$$a_{ij} = c_{ij} + e_{ij}$$

while
$$a_{ji} = c_{ji} + e_{ji} \tag{1.20}$$

On equating the corresponding imaginary parts we obtain

$$b_{ij} = d_{ij} + f_{ij}$$

and
$$b_{ji} = d_{ji} + f_{ji} \tag{1.21}$$

However, one of the matrices is a hermitian matrix and so

$$c_{ij} = c_{ji} \quad \text{and} \quad d_{ij} = -d_{ji}$$

The other matrix is an anti-hermitian matrix which means that

$$e_{ij} = -e_{ji} \quad \text{and} \quad f_{ij} = f_{ji}$$

Equations (1.20) and (1.21) can therefore be rewritten as

$$a_{ij} = c_{ij} + e_{ij}$$
$$a_{ji} = c_{ij} - e_{ij}$$
$$b_{ij} = d_{ij} + f_{ij}$$
$$b_{ji} = -d_{ij} + f_{ij}$$

Solving these four equations gives

$$c_{ij} = \frac{a_{ij} + a_{ji}}{2} \qquad d_{ij} = \frac{b_{ij} - b_{ji}}{2}$$

$$e_{ij} = \frac{a_{ij} - a_{ji}}{2} \qquad f_{ij} = \frac{b_{ij} + b_{ji}}{2}$$

Thus, as the elements of the general square matrix, a_{ij}, a_{ji}, b_{ij} and b_{ji}, are all given, a complete determination of the elements of a hermitian matrix and an anti-hermitian matrix can be made such that their sum is equal to the given square matrix.

Example 1.7:
Express the matrix $\mathbf{A} = \begin{pmatrix} 2+i & 6+4i \\ 2+6i & 3+2i \end{pmatrix}$ as the sum of a hermitian matrix and an anti-hermitian matrix.

Letting the general elements of the hermitian matrix be $c_{ij} + id_{ij}$ and the general element of the anti-hermitian matrix be $e_{ij} + if_{ij}$, we have from the last section

$$c_{11} = \frac{2a_{11}}{2} = 2,\ c_{12} = \frac{a_{12} + a_{21}}{2} = \frac{b+2}{2} = 4 = c_{21},\ c_{22} = \frac{2a_{22}}{2} = 3$$

$$d_{11} = 0,\ d_{12} = \frac{b_{12} - b_{21}}{2} = \frac{4-6}{2} = -1 = -d_{21},\ d_{22} = 0$$

$$e_{11} = 0,\ e_{12} = \frac{a_{12} - a_{21}}{2} = \frac{6-2}{2} = 2 = -e_{21},\ e_{22} = 0$$

$$f_{11} = \frac{2b_{11}}{2} = 1,\ f_{12} = \frac{b_{12} + b_{21}}{2} = \frac{4+6}{2} = 5 = f_{21},\ f_{22} = \frac{2b_{12}}{2} = 2$$

Writing in matrix form we have

$$\begin{pmatrix} 2+\mathrm{i} & 6+4i \\ 2+6i & 3+2i \end{pmatrix} = \begin{pmatrix} 2 & 4-i \\ 4+i & 3 \end{pmatrix} + \begin{pmatrix} i & 2+5i \\ -2+5i & 2i \end{pmatrix}$$

(c) *Any square matrix can be expressed as the sum of an upper triangular matrix, a lower triangular matrix and a diagonal matrix*

This is a fairly obvious combination of matrices and is the least useful of the combinations included in this list. Since the only elements in the triangular section above the diagonal that are non-zero belong to the upper triangular matrix, then these elements must be equal to the corresponding elements in the given matrix. That is, if **A** is the given matrix and **U** the upper triangular matrix,

$$u_{ij} = a_{ij} \text{ for all } j > i$$

Similarly as the only non-zero elements in the combination below the diagonal belong to the lower triangular matrix, they must equal their corresponding elements in the given matrix, that is

$$l_{ij} = a_{ij} \text{ for all } i > j$$

where **L** denotes the lower triangular matrix. This leaves only the diagonal elements for consideration. Denoting the diagonal matrix by Λ, it is required that

$$a_{ij} = u_{ij} + l_{ij} + \lambda_{ij} \text{ for all } i = j \tag{1.22}$$

Since this equation can be satisfied in an infinite number of ways, the stated combination is possible. There are two ways of satisfying equation (1.22) that are more useful in practice:

(i) $\lambda_{ij} = 0$ and either u_{ij} or $l_{ij} = 1$, for all $i = j$.
This means that the diagonal matrix is not used and that either a unit upper triangular matrix or a unit lower triangular matrix is used;

(ii) $u_{ij} = 1$ and $l_{ij} = 1$ for all $i = j$. Thus a unit upper triangular matrix and a unit lower triangular matrix is used.

Example 1.8:

Express $\begin{pmatrix} 4 & 5 & 6 \\ 8 & 9 & 10 \\ 11 & 2 & 19 \end{pmatrix}$ as one of the possible combinations

in section (i) above.

$$\begin{pmatrix} 4 & 5 & 6 \\ 8 & 9 & 10 \\ 11 & 2 & 19 \end{pmatrix} = \begin{pmatrix} 1 & 5 & 6 \\ 0 & 1 & 10 \\ 0 & 0 & 1 \end{pmatrix} + \begin{pmatrix} 3 & 0 & 0 \\ 8 & 8 & 0 \\ 11 & 2 & 18 \end{pmatrix}$$

expressing the given matrix without using a diagonal matrix. Also

$$\begin{pmatrix} 4 & 5 & 6 \\ 8 & 9 & 10 \\ 11 & 2 & 19 \end{pmatrix} = \begin{pmatrix} 1 & 5 & 6 \\ 0 & 1 & 10 \\ 0 & 0 & 1 \end{pmatrix} + \begin{pmatrix} 1 & 0 & 0 \\ 8 & 1 & 0 \\ 11 & 2 & 1 \end{pmatrix} + \begin{pmatrix} 1 & 0 & 0 \\ 0 & 7 & 0 \\ 0 & 0 & 17 \end{pmatrix}$$

expressing the given matrix using both a unit upper triangular matrix and a unit lower triangular matrix.

(d) *Any square matrix can be represented by the product of a unit lower triangular matrix, and an upper triangular matrix*

Let the square matrix **A** have general element a_{ij} while the unit lower triangular matrix **L** has general element l_{ij} and the upper triangular matrix **U** has general element u_{ij}. It is required to determine the elements l_{ij} and u_{ij} such that

$$\mathbf{A} = \mathbf{L}\,\mathbf{U}$$

or

$$a_{ij} = \sum_k l_{ik}\,u_{kj} \tag{1.23}$$

However, we have the restrictions that

$$\begin{aligned} u_{ij} &= 0 \text{ for all } i > j \\ l_{ij} &= 0 \text{ for all } i < j \\ \text{and} \quad l_{ij} &= 1 \text{ for all } i = j \end{aligned} \tag{1.24}$$

Equation (1.23) when $i = 1$ therefore gives

$$a_{1j} = l_{11}\,u_{1j}$$

since $l_{ij} = 0$ for all $j > 1$. However, $l_{11} = 1$, thus

$$u_{1j} = a_{1j}$$

and the first row of the upper triangular matrix can be determined. Consider now equation (1.23) when $j = 1$. It gives

$$a_{i1} = l_{i1}\,u_{11}$$

since $u_{ij} = 0$ for all $i > 1$. As u_{11} has already been determined, this equation is

$$l_{i1} = a_{i1} \,/\, u_{11},$$

allowing a determination of the first column of the lower triangular matrix. With $i = 2$ equation (1.23) now gives

$$a_{2j} = l_{21}\,u_{1j} + l_{22}\,u_{2j}$$

which allows a determination of the second row of the upper triangular matrix, u_{2j}, as l_{21}, u_{1j} and l_{22} are all now known. Repeating this process of considering increasing integral values of i and j in turn all the elements of both the lower triangular matrix and the upper triangular matrix can be determined.

Example 1.9:

Express $\begin{pmatrix} 1 & 2 & 4 \\ 3 & 8 & 14 \\ 2 & 6 & 13 \end{pmatrix}$ as the product of a unit lower triangular matrix and an upper triangular matrix.

Let the lower and upper triangular matrices be given respectively by

$$\begin{pmatrix} 1 & 0 & 0 \\ l_{21} & 1 & 0 \\ l_{31} & l_{32} & 1 \end{pmatrix} \text{ and } \begin{pmatrix} u_{11} & u_{12} & u_{13} \\ 0 & u_{22} & u_{23} \\ 0 & 0 & u_{33} \end{pmatrix}$$

so that

$$\begin{pmatrix} 1 & 2 & 4 \\ 3 & 8 & 14 \\ 2 & 6 & 13 \end{pmatrix} = \begin{pmatrix} 1 & 0 & 0 \\ l_{21} & 1 & 0 \\ l_{31} & l_{32} & 1 \end{pmatrix} \begin{pmatrix} u_{11} & u_{12} & u_{13} \\ 0 & u_{22} & u_{23} \\ 0 & 0 & u_{33} \end{pmatrix}$$

Thus, by considering the first row (i=1) of the product, we obtain

$$u_{11} = 1, u_{12} = 2, u_{13} = 4$$

Considering the first column (j=1) of the product gives

$$l_{21} = 3/u_{11} = 3 \text{ and } l_{31} = 2/u_{11} = 2$$

The second row (i=2) of the product gives

$$8 = l_{21} u_{12} + u_{22} \text{ or } u_{22} = 2$$

and

$$14 = l_{21} u_{13} + u_{23} \text{ or } u_{23} = 2$$

The remaining term of the second column (j=3) in the product gives

$$6 = l_{31} \quad u_{12} + l_{32} \quad u_{22} \text{ or } l_{32} = 1$$

Finally the last row (i=3) or column (j=3) in the product gives

$$13 = l_{31} u_{13} + l_{32} u_{23} + u_{33} \text{ or } u_{33} = 3$$

Rewriting as matrices we have

$$\mathbf{L} = \begin{pmatrix} 1 & 0 & 0 \\ 3 & 1 & 0 \\ 2 & 1 & 1 \end{pmatrix} \text{ and } \mathbf{U} = \begin{pmatrix} 1 & 2 & 4 \\ 0 & 2 & 2 \\ 0 & 0 & 3 \end{pmatrix}$$

where $\mathbf{L}\,\mathbf{U} = \mathbf{A}$.

It is also possible to express a square matrix in terms of the product of a lower triangular matrix and a unit upper triangular matrix. Employing the same notation as in the last discussion equation (1.23) remains valid. The equivalent conditions to conditions (1.24) however are

$$\begin{aligned} u_{ij} &= 0 \text{ for all } i > j \\ l_{ij} &= 0 \text{ for all } i < j \\ u_{ij} &= 0 \text{ for all } i = j \end{aligned} \tag{1.25}$$

Writing equation (1.23) for the case when $j = 1$ therefore gives

$$a_{i1} = l_{i1}\, u_{11}$$

or as $u_{11} = 1$, giving a determination of l_{i1} the first column of the lower triangular matrix. Considering $i = 1$ then gives the first row of the upper triangular matrix and proceeding as in the previous case by alternately selecting increasing values for j and i allows a full determination of both matrices.

This method for representing a square matrix is particularly useful if the matrix is a banded matrix. Using the same notation again, we have

$$a_{ij} = \sum_k l_{ik}\, u_{kj}$$

with the restrictions that

$$\begin{aligned} u_{ij} &= 0 \text{ for all } i > j \\ l_{ij} &= 0 \text{ for all } i < j \\ l_{ij} &= 1 \text{ for all } i = j \\ a_{ij} &= 0 \text{ for all } |i - j| > m \end{aligned} \tag{1.26}$$

Considering the case when $i = 1$ now gives

$$\begin{aligned} u_{ij} &= a_{ij} \text{ for all } j \leq m + 1 \\ u_{ij} &= 0 \quad \text{for all } j > m + 1 \end{aligned}$$

while the case $j = 1$ gives

$$l_{i1} = a_{i1} / u_{11} \text{ for all } i \leq m + 1$$
$$l_{i1} = 0 \qquad \text{for all } i > m + 1$$

No non-zero elements in the first row or column of the triangular matrices have strayed outside the non-zero band in the given matrix. Repetition for higher rows and columns shows that this is the case for all rows and columns.

Example 1.10:

Express the matrix $\mathbf{A} = \begin{pmatrix} 1 & 2 & 0 & 0 \\ 2 & 6 & 2 & 0 \\ 0 & 6 & 9 & 3 \\ 0 & 0 & 12 & 16 \end{pmatrix}$ as the product of

a unit lower triangular matrix and an upper triangular matrix. Letting the lower triangular matrix have elements l_{ij} and the upper triangular matrix have elements u_{ij}, for the first row of $\mathbf{A}$ we have

$$u_{11} = 1, u_{12} = 2, u_{13} = 0, u_{14} = 0$$

Considering the first column of $\mathbf{A}$ gives

$$l_{21} = 2/u_{11} = 2, l_{31} = 0/u_{11} = 0, l_{41} = 0/u_{11} = 0$$

Considering the second row of $\mathbf{A}$ gives

$$u_{22} = 2, u_{23} = 2 \text{ and } u_{24} = 0$$

while the second column gives $l_{32} = 3$ and $l_{42} = 0$. The third row and column give

$$u_{33} = 3, u_{34} = 3 \text{ and } l_{43} = 4$$

Finally the fourth row gives $u_{44} = 4$

Rewriting in matrix form, we have

$$\begin{pmatrix} 1 & 2 & 0 & 0 \\ 2 & 6 & 2 & 0 \\ 0 & 6 & 9 & 3 \\ 0 & 0 & 12 & 16 \end{pmatrix} = \begin{pmatrix} 1 & 0 & 0 & 0 \\ 2 & 1 & 0 & 0 \\ 0 & 3 & 1 & 0 \\ 0 & 0 & 4 & 1 \end{pmatrix} \begin{pmatrix} 1 & 2 & 0 & 0 \\ 0 & 2 & 2 & 0 \\ 0 & 0 & 3 & 3 \\ 0 & 0 & 0 & 4 \end{pmatrix}$$

EXERCISES 1

1. $\mathbf{A}$, $\mathbf{B}$, $\mathbf{C}$ and $\mathbf{D}$ are the four matrices given by

$$\mathbf{A} = \begin{pmatrix} 1 & 1 & 1 \\ 1 & 2 & 3 \\ 1 & 2 & 2 \end{pmatrix}, \ \mathbf{B} = \begin{pmatrix} 0 & 1 & 1 \\ 0 & 1 & 0 \\ 1 & 0 & 0 \end{pmatrix}, \ \mathbf{C} = (0\ 1\ 1), \ \mathbf{D} = \begin{pmatrix} 1 \\ 0 \\ 1 \end{pmatrix}$$

If they exist, calculate the following products:

$\mathbf{A\,B}$, $\mathbf{A\,D}$, $\mathbf{B\,C}$, $\mathbf{C\,A}$, $\mathbf{D\,C}$ and $\mathbf{C\,D}$.

2. $\mathbf{A}$ is the square matrix given by

$$\mathbf{A} = \begin{pmatrix} 0 & 0 & 1 \\ 0 & 1 & 0 \\ 1 & 0 & 0 \end{pmatrix}$$

Show that $\mathbf{A}^3 + 3\mathbf{A}^2 - \mathbf{A} - 3\mathbf{I} = 0$.

3. Given the two matrices $\mathbf{A}$ and $\mathbf{B}$ where

$$\mathbf{A} = \begin{pmatrix} 1 & 2 & 3 \\ 7 & 9 & 11 \\ 2 & 4 & 1 \end{pmatrix} \text{ and } \mathbf{B} = \begin{pmatrix} 2 & 0 & 2 \\ 7 & 3 & 5 \\ 2 & 1 & 8 \end{pmatrix}$$

verify that $(\mathbf{A\,B})^T = \mathbf{B}^T\,\mathbf{A}^T$.

4. With the two matrices

$$\mathbf{A} = \begin{pmatrix} 1 & 2 & 4 \\ 1 & 3 & 4 \\ 2 & 7 & 3 \end{pmatrix} \text{ and } \mathbf{B} = \begin{pmatrix} 1 & 1 & 1 \\ 2 & 4 & 5 \\ 5 & 7 & 8 \end{pmatrix}$$

show that $|\mathbf{A\,B}| = |\mathbf{A}|\,|\mathbf{B}|$.

5. Given

$$\mathbf{A} = \begin{pmatrix} 1 & 7 & 9 \\ 2 & 0 & 2 \\ 3 & 8 & 1 \end{pmatrix}, \ \mathbf{B} = \begin{pmatrix} 2 & 5 & 2 \\ 4 & 3 & 1 \\ 7 & 2 & 1 \end{pmatrix}, \text{ and } \mathbf{C} = \begin{pmatrix} 2 & 5 & 7 \\ 3 & 0 & 1 \\ 2 & 2 & 4 \end{pmatrix}$$

show that $\mathbf{A}(\mathbf{B} + \mathbf{C}) = \mathbf{A\,B} + \mathbf{A\,C}$.

6. Express the matrix $\begin{pmatrix} 1 & 9 & 1 \\ 3 & 2 & 6 \\ 9 & 2 & 3 \end{pmatrix}$ as the sum of a symmetric matrix and an anti-symmetric matrix.

7. Express the matrix $\begin{pmatrix} 2+i & 3-i & 4+2i \\ 5-3i & 3-3i & 1+i \\ 6+4i & 1-i & 6+2i \end{pmatrix}$ as the sum of a hermitian matrix and an anti-hermitian matrix.

8. Express $\begin{pmatrix} 2 & 7 & 9 \\ 6 & 25 & 35 \\ 4 & 30 & 52 \end{pmatrix}$ as the product of a unit lower triangular matrix and an upper triangular matrix.

9. Express $\begin{pmatrix} 2 & 6 & 12 \\ 5 & 18 & 36 \\ 2 & 7 & 18 \end{pmatrix}$ as the product of a lower triangular matrix and a unit upper triangular matrix.

10. Express $\begin{pmatrix} 2 & 3 & 0 \\ 6 & 13 & 6 \\ 0 & 20 & 37 \end{pmatrix}$ as the product of a unit lower triangular matrix and an upper triangular matrix.

2

THE INVERSE OF A MATRIX

2.1. The existence of an inverse matrix

The inverse of the matrix $\mathbf{A}$ has already been defined in section 1.7 to be a matrix $\mathbf{A}^{-1}$ which satisfies

$$\mathbf{A}\,\mathbf{A}^{-1} = \mathbf{I} = \mathbf{A}^{-1}\,\mathbf{A} \tag{2.1}$$

If the rules of multiplication are applicable to all parts of this equation, then the number of rows in $\mathbf{A}$ must equal the number of rows in $\mathbf{I}$, which must in turn be equal to the number of rows in $\mathbf{A}^{-1}$. Similarly, the number of columns in $\mathbf{A}^{-1}$ must equal the number of columns in $\mathbf{I}$ which in turn must equal the number of columns in $\mathbf{A}$. Finally, the number of columns in $\mathbf{A}$ must equal the number of rows in $\mathbf{A}^{-1}$ for the operation $\mathbf{A}\,\mathbf{A}^{-1}$ to be meaningful. All these conditions can be satisfied only if both $\mathbf{A}$ and $\mathbf{A}^{-1}$ are square matrices of the same order. We conclude therefore that an inverse matrix can only exist for a square matrix.

A similar conclusion to the above can be reached by considering a set of linear equations of the form

$$\mathbf{A}\,\mathbf{X} = \mathbf{B} \tag{2.2}$$

where $\mathbf{X}$ and $\mathbf{B}$ are both column vectors of the same order. A solution to these equations is given by

$$\mathbf{X} = \mathbf{A}^{-1}\mathbf{B} \tag{2.3}$$

provided that $\mathbf{A}^{-1}$ exists. The existence of the inverse matrix therefore implies the existence of a solution to the equations. In general a unique solution to equations (2.2) will only exist if the number of equations is equal to the number of unknowns, which of course implies that **A** is a square matrix.

A necessary condition for the existence of an inverse matrix is therefore that the matrix is a square matrix.

Taking the determinant of all the quantities in equation (2.1), which is now possible as all the matrices are square if the equation is meaningful, gives

$$|\mathbf{A}\,\mathbf{A}^{-1}| = |\mathbf{A}|\,|\mathbf{A}^{-1}| = |\mathbf{I}| = 1 \tag{2.4}$$

on making use of the results in section 1.3. A finite inverse will not therefore exist if $|\mathbf{A}| = 0$ and so a further necessary condition for the existence of the inverse matrix $\mathbf{A}^{-1}$ is that **A** is a non-singular matrix, in which case the matrix $\mathbf{A}^{-1}$ will also be non-singular. It has not been shown that these conditions are sufficient, so that an inverse matrix will always exist if the original matrix is square and non-singular, and this will not be done until section 2.4 because it is necessary to define other terms before it becomes possible to do so.

2.2. Minors and cofactors

Let the matrix **A** be of order n and from it delete both the row and the column which contains the element a_{ij} (that is row i and column j). The result is a matrix of order $n - 1$ and the determinant of this matrix is called the MINOR of the element a_{ij}. This minor will be denoted by $\mathcal{A}_{ij}$. As there exists a minor corresponding to every element a_{ij} of the original matrix, then from the matrix **A**, n^2 minors can be calulated.

The COFACTOR corresponding to the element a_{ij} is denoted by A_{ij} and is defined by

$$A_{ij} = (-1)^{i+j}\mathcal{A}_{ij} \tag{2.5}$$

The following result, which will be required in section 2.3, is easily deducable from equation (2.5) and the general rules for the expansion of a determinant:

$$\begin{aligned}\sum_{j=1}^{n} a_{ij}\;A_{kj} &= |\mathbf{A}| \quad i = k \\ &= 0 \quad i \neq k\end{aligned} \tag{2.6}$$

When $i = k$ expression (2.6) is clearly nothing more than the usual expansion of a determinant by row i. If $i \neq k$ then the expression again corresponds to the expansion of a determinant. However, in this determinant row i is included twice (as row k is now the row that has been excluded from A_{kj}) and so by a standard result the determinant must be zero.

2.3. The adjoint matrix

The ADJOINT matrix is defined to be the transpose of the matrix obtained by replacing each element of the matrix $\mathbf{A}$, a_{ij}, by its corresponding cofactor A_{ij}. Thus, if the adjoint matrix is denoted by $\mathbf{AdjA}$ and its general element by $(\mathbf{AdjA})_{ij}$, then

$$(\mathbf{AdjA})_{ij} = A_{ji} \tag{2.7}$$

The adjoint matrix is sometimes called the adjugate matrix.

Consider now the product of a matrix and its corresponding adjoint matrix, that is

$$\mathbf{A}\ \mathbf{AdjA} = \mathbf{B} \tag{2.8}$$

where $\mathbf{B}$ denotes the product so obtained. Consideration of the multiplication rule gives

$$b_{ij} = \sum_{k=1}^{n} a_{ik}\,(\mathbf{AdjA})_{kj}$$

But equation (2.7) gives

$$(\mathbf{AdjA})_{kj} = A_{jk}$$

and so, on remembering expression (2.6),

$$\begin{aligned} b_{ij} = \sum_{k=1}^{n} a_{ik} A_{jk} &= |\mathbf{A}| \quad i = j \\ &= 0 \quad i \neq j \end{aligned}$$

The matrix $\mathbf{B}$ is therefore a diagonal matrix with all its diagonal elements being $|\mathbf{A}|$, or

$$\mathbf{A}\ \mathbf{AdjA} = \mathbf{B} = |\mathbf{A}|\ \mathbf{I} \tag{2.9}$$

2.4. The inverse matrix

Provided the matrix $\mathbf{A}$ is a square matrix, it is always possible to calculate all the elements of the adjoint matrix and so the product $\mathbf{A}\ \mathbf{AdjA}$ can always be calculated. If $\mathbf{A}$ is a non-singular matrix then $|\mathbf{A}|$ is non-zero and it is possible to calculate

$$\frac{\mathbf{A}\ \mathbf{Adj}\ \mathbf{A}}{|\mathbf{A}|}$$

However, from equation (2.8) it can be seen that

$$\frac{\mathbf{A}\ \mathbf{Adj}\ \mathbf{A}}{|\mathbf{A}|} = \mathbf{I}$$

By an exactly similar argument to the above it is possible to show that

$$\frac{\mathbf{Adj}\ \mathbf{A}\ \mathbf{A}}{|\mathbf{A}|} = 1$$

By comparison with equation (2.1) it can therefore be seen that

$$\frac{\mathbf{Adj}\ \mathbf{A}}{|\mathbf{A}|}$$

satisfies all the conditions which the inverse matrix $\mathbf{A}^{-1}$ has to satisfy and therefore it can be identified as the inverse matrix. It has been shown that an inverse matrix can always be found for a square non-singular matrix. In showing that this is the case, a method has also been given which allows a determination of the elements of the inverse matrix.

Example 2.1: Calculate $\mathbf{A}^{-1}$ when $\mathbf{A}$ is given by

$$\mathbf{A} = \begin{pmatrix} 2 & -1 \\ 5 & 3 \end{pmatrix}$$

$$\mathbf{A} = \begin{vmatrix} 2 & -1 \\ 5 & 3 \end{vmatrix} = 11$$

The matrix $\mathbf{A}$ is therefore square and non-singular and so the inverse matrix can be calculated. The matrix of minors is

$$\begin{pmatrix} 3 & 5 \\ -1 & 2 \end{pmatrix}$$

The matrix of cofactors therefore is

$$\begin{pmatrix} 3 & -5 \\ 1 & 2 \end{pmatrix}$$

while the adjoint matrix is

$$\begin{pmatrix} 3 & 1 \\ -5 & 2 \end{pmatrix}$$

Thus,

$$\mathbf{A}^{-1} = \frac{1}{11}\begin{pmatrix} 3 & 1 \\ -5 & 2 \end{pmatrix}$$

It is very easy to verify that $\mathbf{A}\,\mathbf{A}^{-1} = \mathbf{I}$

Example 2.2: Calculate $\mathbf{A}^{-1}$ when $\mathbf{A}$ is given by

$$\mathbf{A} = \begin{pmatrix} 1 & 2 & 3 \\ 2 & -1 & 4 \\ 0 & -1 & 1 \end{pmatrix}$$

$$|\mathbf{A}| = \begin{vmatrix} 1 & 2 & 3 \\ 2 & -1 & 4 \\ 0 & -1 & 1 \end{vmatrix} = \begin{vmatrix} 1 & 5 & 3 \\ 2 & 3 & 4 \\ 0 & 0 & 1 \end{vmatrix} = -7$$

The cofactors are given by $(-1)^{i+j}\,a_{ij}$. Calculating these gives

$$A_{11} = (-1)^2 \begin{vmatrix} -1 & 4 \\ -1 & 1 \end{vmatrix} = 3, \quad A_{12} = (-1)^3 \begin{vmatrix} 2 & 4 \\ 0 & 1 \end{vmatrix} = -2,$$

$$A_{13} = (-1)^4 \begin{vmatrix} 2 & -1 \\ 0 & -1 \end{vmatrix} = -2$$

$$A_{21} = (-1)^3 \begin{vmatrix} 1 & 3 \\ -1 & 1 \end{vmatrix} = -5, \quad A_{22} = (-1)^4 \begin{vmatrix} 1 & 3 \\ 0 & 1 \end{vmatrix} = -1,$$

$$A_{23} = (-1)^5 \begin{vmatrix} 1 & 2 \\ 0 & -1 \end{vmatrix} = 1$$

$$A_{31} = (-1)^4 \begin{vmatrix} 2 & 3 \\ -1 & 4 \end{vmatrix} = 11, \quad A_{32} = (1-)^5 \begin{vmatrix} 1 & 3 \\ 2 & 4 \end{vmatrix} = 2,$$

$$A_{33} = (-1)^6 \begin{vmatrix} 1 & 2 \\ 2 & -1 \end{vmatrix} = -5$$

$$\textbf{Adj A} = \begin{pmatrix} 3 & -2 & -2 \\ -5 & 1 & 1 \\ 11 & 2 & -5 \end{pmatrix}^T = \begin{pmatrix} 3 & -5 & 11 \\ -2 & 1 & 2 \\ -2 & 1 & -5 \end{pmatrix}$$

and so

$$\mathbf{A}^{-1} = \frac{1}{7}\begin{pmatrix} 3 & -5 & 11 \\ -2 & 1 & 2 \\ -2 & 1 & -5 \end{pmatrix}$$

We note that considerably more arithmetic manipulation is involved in this calculation for a matrix of order 3 than was the case in the calculation for a matrix of order 2 as was carried out in example 2.1.

2.5. The inverse of the transpose of a matrix

Let $\mathbf{A}$ be a square non-singular matrix so that its inverse, $\mathbf{A}^{-1}$, exists and satisfies the equation

$$\mathbf{A}\,\mathbf{A}^{-1} = \mathbf{I} = \mathbf{A}^{-1}\,\mathbf{A}$$

Taking the transpose of all terms of this equation yields

$$(\mathbf{A}^{-1})^T\,\mathbf{A}^T = \mathbf{I} = \mathbf{A}^T\,(\mathbf{A}^{-1})^T \tag{2.10}$$

on using the rule for obtaining the transpose of a product. This equation (2.10) shows that the inverse of the matrix $\mathbf{A}^T$ is the matrix $(\mathbf{A}^{-1})^T$ since it satisfies the equation which the inverse must satisfy. It therefore follows that

$$(\mathbf{A}^{-1})^T = (\mathbf{A}^T)^{-1}$$

and it has been shown that the inverse of the transpose of a matrix is equal to the transpose of the inverse matrix.

2.6. The inverse of the product of matrices

Let $\mathbf{A}$ and $\mathbf{B}$ be two square non-singular matrices of the same order. As $|\mathbf{A}| \neq 0$ and $|\mathbf{B}| \neq 0$, both the inverse matrices, $\mathbf{A}^{-1}$ and $\mathbf{B}^{-1}$, exist. As

$$\mathbf{B}\,\mathbf{B}^{-1} = \mathbf{I}$$

then

$$\mathbf{A}\,\mathbf{B}\,\mathbf{B}^{-1}\,\mathbf{A}^{-1} = \mathbf{A}\,\mathbf{I}\,\mathbf{A}^{-1} = \mathbf{A}\,\mathbf{A}^{-1} = \mathbf{I}$$

or

$$(\mathbf{A}\,\mathbf{B})\,(\mathbf{B}^{-1}\,\mathbf{A}^{-1}) = \mathbf{I}$$

which shows that the inverse of the product $\mathbf{A}\,\mathbf{B}$ is $\mathbf{B}^{-1}\,\mathbf{A}^{-1}$. This procedure can obviously be extended to cover any number of matrices of the same order and so the inverse of

$$\mathbf{A}_1\,\mathbf{A}_2\,\mathbf{A}_3 \ldots \mathbf{A}_n \text{ is } \mathbf{A}_n^{-1}\quad \mathbf{A}_{n-1}^{-1} \ldots\ \mathbf{A}_2^{-1}\,\mathbf{A}_1^{-1} \qquad (2.11)$$

2.7. The Gauss–Jordan method for calculating the inverse matrix

It has already been pointed out that the amount of manipulative work involved in calculating the inverse of a matrix of order 3 is very much greater than the corresponding amount for a matrix of order 2. In fact, evaluating the inverse of a matrix of order n by the method given in section 2.4 involves evaluating one determinant of order n and n^2 determinants of order n–1 (that is, all the cofactors). The number of multiplications involved in evaluating a determinant of order m is about $m!$ and so the total number of multiplications that have to be carried out in order to evaluate the inverse of a matrix of order n is

$$n! + (n - 1)!n^2 = (n + 1)!$$

Now 18! is about 6.4×10^{15} and so the fastest electronic computers, having a multiplication time of about 10^{-6} seconds, will take 6.4×10^9 seconds or over 200 years to calculate the inverse of a matrix of order 17. In some physical problems, matrices of order greater than 17 have to be used and so it is essential to use some other method to evaluate the inverse of such a matrix. One of the simplest and most efficient methods that has been devised is the method called Gauss–Jordan elimination. Historically this method is derived from the Gaussian elimination method for solving equations which will be described in Chapter 3, but it is more appropriate to discuss methods for finding inverses before methods for solving linear equations.

Let $\mathbf{A}$ denote a square non-singular matrix. Then a set of linear equations of the form

$$\mathbf{A}\,\mathbf{X} = \mathbf{Y}$$

can equally be written as

$$\mathbf{A}\,\mathbf{X} = \mathbf{I}\,\mathbf{Y} \qquad (2.12)$$

while its solution will be

$$\mathbf{I}\,\mathbf{X} = \mathbf{A}^{-1}\,\mathbf{Y} \qquad (2.13)$$

However, the set of equations (2.12) can be solved by successively eliminating coefficients through substracting a suitable multiple of the first equation from the remainder and then repeating with successive equations. The result is clearly to generate equation (2.13). Now, the values of $\mathbf{Y}$ and $\mathbf{X}$ play no real part in this elimination; the important point is that the matrices $\mathbf{A}$ and $\mathbf{I}$, by successive elimination by the rows of $\mathbf{A}$, are reduced to the matrices $\mathbf{I}$ and $\mathbf{A}^{-1}$. It therefore gives a method for evaluating $\mathbf{A}^{-1}$.

Example 2.3: By the Gauss–Jordan elimination method, evaluate the inverse of $\mathbf{A}$ where

$$\mathbf{A} = \begin{pmatrix} 1 & 2 & 3 \\ 2 & -1 & 4 \\ 0 & -1 & 1 \end{pmatrix}$$

An easier way to set out the essential features of the equation corresponding to equation (2.12) is as follows:

$$\begin{array}{rrr|rrr} 1 & 2 & 3 & 1 & 0 & 0 \\ 2 & -1 & 4 & 0 & 1 & 0 \\ 0 & -1 & 1 & 0 & 0 & 1 \end{array}$$

Using R_i to denote the rows under discussion, the following procedure then yields the inverse.

$R_2 \longrightarrow R_2 - 2R_1$

$$\begin{array}{rrr|rrr} 1 & 2 & 3 & 1 & 0 & 0 \\ 0 & -5 & -2 & -2 & 1 & 0 \\ 0 & -1 & 1 & 0 & 0 & 1 \end{array}$$

$R_3 \longrightarrow 5R_3 - R_2$

$$\begin{array}{rrr|rrr} 1 & 2 & 3 & 1 & 0 & 0 \\ 0 & -5 & -2 & -2 & 1 & 0 \\ 0 & 0 & 7 & 2 & -1 & 5 \end{array}$$

$R_1 \longrightarrow 7R_1 - 3R_3$; $R_2 \longrightarrow 7R_2 + 2R_3$

$$\begin{array}{rrr|rrr} 7 & 14 & 0 & 1 & 3 & -15 \\ 0 & -35 & 0 & -10 & 5 & 10 \\ 0 & 0 & 7 & 2 & -1 & 5 \end{array}$$

$R_2 \longrightarrow R_2/-5$

$$\begin{array}{rrr|rrr} 7 & 14 & 0 & 1 & 3 & -15 \\ 0 & 7 & 0 & 2 & -1 & -2 \\ 0 & 0 & 7 & 2 & -1 & 5 \end{array}$$

$R_1 \longrightarrow R_1 - 2R_2$

$$\begin{array}{rrr|rrr} 7 & 0 & 0 & -3 & 8 & -11 \\ 0 & 7 & 0 & 2 & -1 & -2 \\ 0 & 0 & 7 & 2 & -1 & 5 \end{array}$$

Hence the inverse is $\frac{1}{7}\begin{pmatrix} -3 & 5 & -11 \\ 2 & -1 & -2 \\ 2 & -1 & 5 \end{pmatrix}$

as can be verified by referring to example 2.2.

2.8. Pivoting

In the above example all the elements were maintained as integers by considering only operations of the form $R_1 \longrightarrow 7R_1 - 3R_3$. In practice, the computer would carry out the equivalent operation $R_1 \longrightarrow R_1 - \frac{3}{7} R_3$, thus generating non-integer terms: this is immaterial as most matrices that occur in practice have non-integer elements. Unfortunately with non-integer elements round-off errors can occur and they become relatively more important when they occur in conjunction with a small number. Such errors can have very drastic effects if division by a small number is carried out. PIVOTING is an attempt to minimize this type of error by ensuring that when division is carried out, it is always by the largest element available. The procedure is to locate the largest element in the matrix. As re-arranging complete rows now corresponds to re-arranging the order of the equations, it is allowable and so the rows are re-arranged until the largest element occurs on the leading diagonal. This largest element is then used to eliminate all the other elements from the column in which it appears. The whole process is repeated, disregarding elements in rows that have already been used for elimination purposes, until a diagonal matrix is formed. Then simple division produces the unit matrix. (In practice this last division is carried out simultaneously with the elimination.)

Example 2.4: Find the inverse of the matrix **A** where

$$\mathbf{A} = \begin{pmatrix} 1 & 1 & 1 \\ 1 & 3 & 6 \\ 1 & 2 & 3 \end{pmatrix}$$

The same notation, that R_i denotes row i, as was used in example 2.3 is employed. As in that example we shall preserve the elements as integers by multiplying elements in rows, rather than dividing these elements, that is, we will consider operations of the form $aR_i - bR_j$ rather than of the form $R_i - (b/a)R_j$. The largest element in the matrix occurs in row 2 column 3. Row 2 and row 3 are therefore interchanged, so that it now occurs in row 3 column 3, giving

$$\left[\begin{array}{ccc|ccc} 1 & 1 & 1 & 1 & 0 & 0 \\ 1 & 2 & 3 & 0 & 0 & 1 \\ 1 & 3 & 6 & 0 & 1 & 0 \end{array}\right]$$

Note that the rows on the right-hand side of the array have also been interchanged. Eliminating in column 3 using this element gives

$$R_1 \rightarrow 6R_1 - R_3, \quad R_2 \rightarrow 2R_2 - R_3$$

$$\left[\begin{array}{ccc|ccc} 5 & 3 & 0 & 6 & -1 & 1 \\ 1 & 1 & 0 & 0 & -1 & 2 \\ 1 & 3 & 6 & 0 & 1 & 0 \end{array}\right]$$

Row 3 has now been used for elimination purposes and the largest element outside this row has now to be located. This element occurs in row 1 column 1 and so an interchange of rows is unnecessary. Eliminating in column 1 using this element gives

$$R_2 \rightarrow 5R_2 - R_1, \quad R_3 \rightarrow 5R_3 - R_1$$

$$\left[\begin{array}{ccc|ccc} 5 & 3 & 0 & 6 & -1 & 0 \\ 0 & 2 & 0 & -6 & -4 & 10 \\ 0 & 12 & 30 & -6 & 6 & 0 \end{array}\right]$$

Row 2 is now the only row that has not been used for eliminating purposes, and so using it gives

$$R_1 \rightarrow 2R_1 - R_2, \quad R_3 \rightarrow 5R_3 - R_2$$

$$\left[\begin{array}{ccc|ccc} 10 & 0 & 0 & 30 & 10 & -30 \\ 0 & 2 & 0 & -6 & -4 & 10 \\ 0 & 0 & 30 & 30 & 30 & -60 \end{array}\right]$$

Finally

$$R_1 \rightarrow R_1/10, \quad R_2 \rightarrow R_2/2, \quad R_3 \rightarrow R_3/30$$

$$\left[\begin{array}{ccc|ccc} 1 & 0 & 0 & 3 & 1 & -3 \\ 0 & 1 & 0 & -3 & -2 & 5 \\ 0 & 0 & 1 & 1 & 1 & -2 \end{array}\right]$$

and so the inverse of $\mathbf{A}$ is

$$\begin{pmatrix} 3 & 1 & -3 \\ -3 & -2 & 5 \\ 1 & 1 & -2 \end{pmatrix}$$

2.9. Improving an approximate inverse

In some problems, even after pivoting, rounding-off errors may have accumulated and so the inverse that has been found may only be approximate. In other problems, because of the nature of the problem and its expected solution, it is possible that an approximate inverse has been arrived at without evaluation by using any of the standard methods. In either event it is desirable to be able to improve upon the approximate inverse that has been derived.

Let $\mathbf{B}$ be an approximation of the inverse of the square matrix $\mathbf{A}$. Then,

$$\mathbf{A\,B} = \mathbf{I} - \mathbf{E} \tag{2.14}$$

where $\mathbf{E}$ is a matrix which represents the deviation of the product $\mathbf{A\,B}$ from $\mathbf{I}$. Should $\mathbf{B}$ perchance be the exact inverse then $\mathbf{E} = \mathbf{O}$.

Solving equation (2.14) to evaluate $\mathbf{B}$ gives

$$\mathbf{B} = \mathbf{A}^{-1} - \mathbf{A}^{-1}\,\mathbf{E}$$

and on rearranging this equation we have

$$\mathbf{A}^{-1} = \mathbf{B} + \mathbf{A}^{-1}\,\mathbf{E} \tag{2.15}$$

Substituting for $\mathbf{A}^{-1}$ into the right-hand side of this equation gives

$$\mathbf{A}^{-1} = \mathbf{B} + (\mathbf{B} + \mathbf{A}^{-1}\,\mathbf{E})\,\mathbf{E} = \mathbf{B} + \mathbf{B\,E} + \mathbf{A}^{-1}\,\mathbf{E}^2$$

Repeating the substitution for $\mathbf{A}^{-1}$ a sufficient number of times gives

$$\mathbf{A}^{-1} = \mathbf{B} + \mathbf{B\,E} + \mathbf{B\,E}^2 + \ldots\ \mathbf{B\,E}^n + \ldots \tag{2.16}$$

the inclusion of each additional term improving the approximation. Though a good approximation may be obtained in this way a far more efficient way to obtain a good approximation (that is, to obtain the highest possible accuracy with a given number of operations) is to take

$$\mathbf{B} + \mathbf{B}\,\mathbf{E} + \mathbf{B}\,\mathbf{E}^2 = \mathbf{C}, \text{ say,} \tag{2.17}$$

and repeat with $\mathbf{C}$ as a new approximate inverse. If, for example, we allow 12 matrix multiplications only, using equation (2.16) would give as its highest term $\mathbf{B}\,\mathbf{E}^{11}$ (remembering that one multiplication is required to evaluate $\mathbf{E}$). Using the alternative procedure the highest term in the expansion obtained with 12 matrix multiplications is $\mathbf{B}\,\mathbf{E}^{54}$, a very much better level of accuracy.

Example 2.5: $\mathbf{A}$ is the matrix given

$$\mathbf{A} = \begin{pmatrix} 1 & 1 & 1 \\ 1 & 2 & 3 \\ 2 & 3 & 8 \end{pmatrix}$$

and $\mathbf{B}$ is an approximate inverse that has been found, where

$$\mathbf{B} = \begin{pmatrix} 1{\cdot}7 & -1{\cdot}2 & 0{\cdot}2 \\ -0{\cdot}5 & 1{\cdot}5 & -0{\cdot}5 \\ -0{\cdot}2 & -0{\cdot}25 & 0{\cdot}2 \end{pmatrix}$$

Improve the accuracy of the inverse.

$$\mathbf{A}\,\mathbf{B} = \begin{pmatrix} 1 & 1 & 1 \\ 1 & 2 & 3 \\ 2 & 3 & 8 \end{pmatrix} \begin{pmatrix} 1{\cdot}7 & -1{\cdot}2 & 0{\cdot}2 \\ -0{\cdot}5 & 1{\cdot}5 & -0{\cdot}5 \\ -0{\cdot}2 & -0{\cdot}25 & 0{\cdot}2 \end{pmatrix}$$

$$= \begin{pmatrix} 1 & 0{\cdot}05 & -0{\cdot}1 \\ 0{\cdot}1 & 1{\cdot}05 & -0{\cdot}2 \\ 0{\cdot}3 & 0{\cdot}1 & 0{\cdot}5 \end{pmatrix}$$

Hence

$$\mathbf{E} = \mathbf{I} - \mathbf{A}\,\mathbf{B} = -\begin{pmatrix} 0 & 0{\cdot}05 & -0{\cdot}1 \\ 0{\cdot}1 & 0{\cdot}05 & -0{\cdot}2 \\ 0{\cdot}3 & 0{\cdot}1 & -0{\cdot}5 \end{pmatrix}$$

$$\mathbf{B}\,\mathbf{E} = -\begin{pmatrix} 1{\cdot}7 & -1{\cdot}2 & 0{\cdot}2 \\ -0{\cdot}5 & 1{\cdot}5 & -0{\cdot}5 \\ -0{\cdot}2 & -0{\cdot}25 & 0{\cdot}2 \end{pmatrix} \begin{pmatrix} 0 & 0{\cdot}05 & -0{\cdot}1 \\ 0{\cdot}1 & 0{\cdot}05 & -0{\cdot}2 \\ 0{\cdot}3 & 0{\cdot}1 & -0{\cdot}5 \end{pmatrix}$$

$$= \begin{pmatrix} 0{\cdot}06 & -0{\cdot}045 & 0{\cdot}03 \\ 0 & 0 & 0 \\ -0{\cdot}035 & 0{\cdot}0025 & 0{\cdot}03 \end{pmatrix}$$

$$\mathbf{B\,E}^2 = -\begin{pmatrix} 0{\cdot}0045 & 0{\cdot}00375 & -0{\cdot}012 \\ 0 & 0 & 0 \\ 0{\cdot}00925 & 0{\cdot}0014 & -0{\cdot}012 \end{pmatrix}$$

$$\mathbf{B} + \mathbf{B\,E} + \mathbf{B\,E}^2 = \begin{pmatrix} 1{\cdot}755 & -1{\cdot}249 & 0{\cdot}242 \\ -0{\cdot}5 & 1{\cdot}5 & -0{\cdot}5 \\ -0{\cdot}244 & -0{\cdot}249 & 0{\cdot}242 \end{pmatrix}$$

Repetition of the procedure will lead to further improvement. The inverse of the original matrix can in fact be obtained fairly easily and the inverse should be

$$\mathbf{A} = \begin{pmatrix} 1{\cdot}75 & -1{\cdot}25 & 0{\cdot}25 \\ -0{\cdot}5 & 1{\cdot}5 & -0{\cdot}5 \\ -0{\cdot}25 & -0{\cdot}25 & 0{\cdot}25 \end{pmatrix}$$

2.10 Partitioned matrix

For certain types of calculation, in particular the calculation of the inverse of a matrix with a block of zeros in it, considerable simplification is possible if the matrix is subdivided into a number of submatrices and each submatrix then regarded as matrix elements. Consider first the multiplication of a matrix and a vector. Let $\mathbf{A}$ be a matrix of order $m \times n$ and divide this into two submatrices, one of order $m \times q$ and the other of order $m \times n{-}q$, denoted by $\mathbf{A}_1$ and $\mathbf{A}_2$. Similarly, subdivide the vector $\mathbf{X}$ into two subvectors, $\mathbf{X}_1$ and $\mathbf{X}$, $\mathbf{X}_1$ of order q and $\mathbf{X}_2$ of order $n{-}q$. Consider then the product

$$\mathbf{A\,X} = (\mathbf{A}_1\,\mathbf{A}_2)\begin{pmatrix} \mathbf{X}_1 \\ \mathbf{X}_2 \end{pmatrix}, \text{ or}$$

$$\begin{pmatrix} a_{11} & \cdots & a_{1q} & a_{1q+1} & \cdots & a_{1n} \\ \cdots & \cdots & \cdots & \cdots & \cdots & \cdots \\ a_{m1} & \cdots & a_{mq} & a_{mq+1} & \cdots & a_{mn} \end{pmatrix} \begin{pmatrix} x_1 \\ \cdot \\ \cdot \\ \cdot \\ x_q \\ x_{q+1} \\ \cdot \\ \cdot \\ \cdot \\ x_n \end{pmatrix}$$

Let the product be a vector $\mathbf{Y}$, then by the rules of multiplication

$$y_i = \sum_{k=1}^{n} a_{ik} x_k = \sum_{k=1}^{q} a_{ik} x_k + \sum_{k=q+1}^{n} a_{ik} x_k \quad (2.18)$$

But the first summation is the ith element in the product $\mathbf{A}_1 \mathbf{X}_1$ while the second summation is the ith element in the product $\mathbf{A}_2 \mathbf{X}_2$ and so

$$\mathbf{A}\,\mathbf{X} = (\mathbf{A}_1\,\mathbf{A}_2)\begin{pmatrix}\mathbf{X}_1\\ \mathbf{X}_2\end{pmatrix} = \mathbf{A}_1\mathbf{X}_1 + \mathbf{A}_2\mathbf{X}_2, \quad (2.19)$$

just as if the submatrices were matrix elements. Now let the matrix $\mathbf{A}$ also be divided between row p and row $p + 1$ so that four submatrices exist, that is

$$\mathbf{A} = \begin{pmatrix}\mathbf{A}_{11} & \mathbf{A}_{12}\\ \mathbf{A}_{21} & \mathbf{A}_{22}\end{pmatrix}$$

The equivalent statement to that expressed in equation (2.18) now becomes

for $\quad i \leqslant p \quad y_i = \sum_{k=1}^{q} a_{ik} x_k + \sum_{k=q+1}^{n} a_{ik} x_k$

for $\quad i > p \quad y_i = \sum_{k=1}^{q} a_{ik} x_k + \sum_{k=q+1}^{n} a_{ik} x_k$

or

$$\begin{pmatrix}\mathbf{A}_{11} & \mathbf{A}_{12}\\ \mathbf{A}_{21} & \mathbf{A}_{22}\end{pmatrix}\begin{pmatrix}\mathbf{X}_1\\ \mathbf{X}_2\end{pmatrix} = \begin{pmatrix}\mathbf{Y}_1\\ \mathbf{Y}_2\end{pmatrix}$$

where

$$\mathbf{Y}_1 = \mathbf{A}_{11}\mathbf{X}_1 + \mathbf{A}_{12}\mathbf{X}_2$$

and

$$\mathbf{Y}_2 = \mathbf{A}_{21}\mathbf{X}_1 + \mathbf{A}_{22}\mathbf{X}_2$$

Again the submatrices behave as if they were matrix elements.

Finally, if the vector $\mathbf{X}$ is replaced by a matrix $\mathbf{B}$ so that the vector $\mathbf{Y}$ also has to be replaced by a matrix $\mathbf{C}$, when all three matrices are partitioned in a suitable matching way into four submatrices, we have

$$\begin{pmatrix}\mathbf{A}_{11} & \mathbf{A}_{12}\\ \mathbf{A}_{21} & \mathbf{A}_{22}\end{pmatrix}\begin{pmatrix}\mathbf{B}_{11} & \mathbf{B}_{12}\\ \mathbf{B}_{21} & \mathbf{B}_{22}\end{pmatrix} = \begin{pmatrix}\mathbf{C}_{11} & \mathbf{C}_{12}\\ \mathbf{C}_{21} & \mathbf{C}_{22}\end{pmatrix}$$

and it follows from exactly similar equations to equation (2.18) that

$$\mathbf{A}_{11}\mathbf{B}_{11} + \mathbf{A}_{12}\mathbf{B}_{21} = \mathbf{C}_{11}$$
$$\mathbf{A}_{11}\mathbf{B}_{12} + \mathbf{A}_{12}\mathbf{B}_{22} = \mathbf{C}_{12}$$

$$\mathbf{A}_{21}\,\mathbf{B}_{11} + \mathbf{A}_{22}\,\mathbf{B}_{21} = \mathbf{C}_{21}$$

$$\mathbf{A}_{21}\,\mathbf{B}_{12} + \mathbf{A}_{22}\,\mathbf{B}_{22} = \mathbf{C}_{22}$$

again as if all the submatrices were elements of a matrix. Hence, provided the partitioning is suitably done, matrices may always be partitioned into submatrices and these submatrices can then be treated like elements of the matrices.

2.11. Calculating the inverse by means of partitioning

If a matrix contains a block of zero elements the work involved in calculating the inverse can be reduced by partitioning. Consider a matrix $\mathbf{A}$ with a block of zeros in one of the corners of the matrix off the leading diagonal which we take to be the top right-hand corner; the argument is the same if the block of zeros occurs in the bottom right-hand corner. This matrix is partitioned into four submatrices in such a way that the submatrix in the top right-hand corner is as large as possible while containing only zero elements. At the same time the two submatrices on the leading diagonal must remain square. The partitioning is therefore such that

$$\mathbf{A} = \begin{pmatrix} \mathbf{A}_{11} & \mathbf{A}_{12} \\ \mathbf{A}_{21} & \mathbf{A}_{22} \end{pmatrix}$$

with $\mathbf{A}_{12} = \mathbf{O}$ and $\mathbf{A}_{11}$ and $\mathbf{A}_{22}$ both square matrices. As $\mathbf{A}_{12} = \mathbf{O}$, $|\mathbf{A}| = |\mathbf{A}_{11}|\ |\mathbf{A}_{22}|$ and so, if $\mathbf{A}$ is non-singular in order that its inverse exists, then both $\mathbf{A}_{11}$ and $\mathbf{A}_{22}$ are non-singular so that $\mathbf{A}_{11}^{-1}$ and $\mathbf{A}_{22}^{-1}$ exist.

Suppose the inverse of $\mathbf{A}$ is given by the matrix $\mathbf{B}$, suitably partitioned such that

$$\mathbf{B} = \begin{pmatrix} \mathbf{B}_{11} & \mathbf{B}_{12} \\ \mathbf{B}_{21} & \mathbf{B}_{22} \end{pmatrix} \quad \text{and} \quad \mathbf{A}\,\mathbf{B} = \mathbf{I}, \text{ or}$$

$$\begin{pmatrix} \mathbf{A}_{11} & \mathbf{A}_{12} \\ \mathbf{A}_{21} & \mathbf{A}_{22} \end{pmatrix} \begin{pmatrix} \mathbf{B}_{11} & \mathbf{B}_{12} \\ \mathbf{B}_{21} & \mathbf{B}_{22} \end{pmatrix} = \begin{pmatrix} \mathbf{I} & \mathbf{O} \\ \mathbf{O} & \mathbf{I} \end{pmatrix} \tag{2.20}$$

Treating each submatrix as a matrix element we have

$$\mathbf{A}_{11}\ \mathbf{B}_{11} + \mathbf{A}_{12}\ \mathbf{B}_{21} = \mathbf{I}$$

or $$\mathbf{B}_{11} = \mathbf{A}_{11}^{-1} \text{ as } \mathbf{A}_{12} = \mathbf{O}$$

Also $$\mathbf{A}_{11}\ \mathbf{B}_{12} + \mathbf{A}_{12}\ \mathbf{B}_{22} = \mathbf{O}$$

or $$\mathbf{B}_{12} = \mathbf{0} \text{ again as } \mathbf{A}_{12} = \mathbf{0}$$

Similarly $$\mathbf{A}_{21}\mathbf{B}_{11} + \mathbf{A}_{22}\mathbf{B}_{21} = \mathbf{0}$$

or $$\mathbf{B}_{21} = -\mathbf{A}_{22}^{-1}\mathbf{A}_{21}\mathbf{A}_{11}^{-1} \text{ as } \mathbf{B}_{11} = \mathbf{A}_{11}^{-1}$$

and $$\mathbf{A}_{21}\mathbf{B}_{12} + \mathbf{A}_{22}\mathbf{B}_{22} = \mathbf{I}$$

or $$\mathbf{B}_{22} = \mathbf{A}_{22}^{-1} \text{ as } \mathbf{B}_{12} = \mathbf{0}$$

Writing in the deduced values for the submatrices

$$\mathbf{A}^{-1} = \begin{pmatrix} \mathbf{A}_{11}^{-1} & \mathbf{0} \\ -\mathbf{A}_{22}^{-1}\mathbf{A}_{21}\mathbf{A}_{11}^{-1} & \mathbf{A}_{22}^{-1} \end{pmatrix}$$

The work in evaluating the inverse is therefore reduced as the inverse of two smaller matrices only has to be evaluated.

Example 2.6: Find the inverse of the matrix $\mathbf{A}$ where

$$\mathbf{A} = \begin{pmatrix} 1 & 1 & 0 & 0 \\ 1 & 2 & 0 & 0 \\ 7 & 8 & 2 & -1 \\ -1 & 1 & -5 & 3 \end{pmatrix}$$

The partitioning to be used in this example is obvious, giving

$$\mathbf{A}_{11} = \begin{pmatrix} 1 & 1 \\ 1 & 2 \end{pmatrix}, \; \mathbf{A}_{12} = \begin{pmatrix} 0 & 0 \\ 0 & 0 \end{pmatrix}, \; \mathbf{A}_{21} = \begin{pmatrix} 7 & 8 \\ -1 & 1 \end{pmatrix},$$

and $$\mathbf{A}_{22} = \begin{pmatrix} 2 & -1 \\ -5 & 3 \end{pmatrix}$$

Making use of the theory gives

$$\mathbf{B}_{11} = \mathbf{A}_{11}^{-1} = \begin{pmatrix} 2 & -1 \\ -1 & 1 \end{pmatrix}, \; \mathbf{B}_{22} = \mathbf{A}_{22}^{-1} = \begin{pmatrix} 3 & 1 \\ 5 & 2 \end{pmatrix},$$

$$\mathbf{B}_{21} = -\mathbf{A}_{22}^{-1}\mathbf{A}_{21}\mathbf{A}_{11}^{-1} = -\begin{pmatrix} 3 & 1 \\ 5 & 2 \end{pmatrix}\begin{pmatrix} 7 & 8 \\ -1 & 1 \end{pmatrix}\begin{pmatrix} 2 & -1 \\ -1 & 1 \end{pmatrix}$$

$$= -\begin{pmatrix} 3 & 1 \\ 5 & 2 \end{pmatrix}\begin{pmatrix} 6 & 1 \\ -3 & 2 \end{pmatrix} = \begin{pmatrix} -15 & -5 \\ -24 & -9 \end{pmatrix}$$

and $$\mathbf{B}_{12} = \mathbf{0}$$

On reassembling the partitioned matrix, $\mathbf{B}$, we obtain

$$\mathbf{B} = \mathbf{A}^{-1} = \begin{pmatrix} 2 & -1 & 0 & 0 \\ -1 & 1 & 0 & 0 \\ -15 & -5 & 3 & 1 \\ -24 & -9 & 5 & 2 \end{pmatrix}$$

2.12. Application of partitioning to linear equations

Although linear equations will not be dealt with until the next chapter this application is so closely connected to the partitioning of matrices that it seems more appropriate to include it at this stage. Consider a matrix equation of the form

$$\mathbf{A}\,\mathbf{X} = \mathbf{B} \tag{2.21}$$

where $\mathbf{B}$ and $\mathbf{X}$ are both vectors of the same order and $\mathbf{A}$ is a square matrix also of the same order. However, instead of the usual situation where all the elements of $\mathbf{B}$ are given, only some of the elements are given and, without loss of generality, we can assume that equation (2.21) has been so written that the known elements all occur as the last elements of $\mathbf{B}$, $\mathbf{B}_2$, say. The corresponding elements $\mathbf{X}_2$ of $\mathbf{X}$ are unknown. The remaining elements of $\mathbf{B}$, $\mathbf{B}_1$, say, are unknown but the corresponding elements of $\mathbf{X}$, $\mathbf{X}_1$ are known. If the matrix $\mathbf{A}$ is suitably partitioned, equation (2.21) can be rewritten as

$$\begin{pmatrix} \mathbf{A}_{11} & \mathbf{A}_{12} \\ \mathbf{A}_{21} & \mathbf{A}_{22} \end{pmatrix} \begin{pmatrix} \mathbf{X}_1 \\ \mathbf{X}_2 \end{pmatrix} = \begin{pmatrix} \mathbf{B}_1 \\ \mathbf{B}_2 \end{pmatrix} \tag{2.22}$$

where $\mathbf{X}_1$ and $\mathbf{B}_2$ are known and because $\mathbf{X}_1$ and $\mathbf{B}_1$ are of the same order, $\mathbf{A}_{11}$ and $\mathbf{A}_{22}$ are both square matrices.

Equation (2.22) implies that

$$\mathbf{A}_{11}\,\mathbf{X}_1 + \mathbf{A}_{12}\,\mathbf{X}_2 = \mathbf{B}_1 \tag{2.23}$$

and

$$\mathbf{A}_{21}\,\mathbf{X}_1 + \mathbf{A}_{22}\,\mathbf{X}_2 = \mathbf{B}_2 \tag{2.24}$$

As $\mathbf{A}_{22}$ is a square matrix and $\mathbf{X}_1$ and $\mathbf{B}_2$ are known, from equation (2.24) we have

$$\mathbf{X}_2 = \mathbf{A}_{22}^{-1}\,(\mathbf{B}_2 - \mathbf{A}_{21}\,\mathbf{X}_1) \tag{2.25}$$

and, on substituting into equation (2.23),

$$\mathbf{A}_{11}\,\mathbf{X}_1 + \mathbf{A}_{12}\,\mathbf{A}_{22}^{-1}\,(\mathbf{B}_2 - \mathbf{A}_{21}\,\mathbf{X}_1) = \mathbf{B}_1,$$

which determines $\mathbf{B}_1$. Both unknown vectors have therefore been evaluated.

Example 2.7: A set of linear equations has the form

$$\begin{pmatrix} 40 & 3 & -10 & -6 \\ 3 & 10 & -3 & -10 \\ -10 & -3 & 10 & 3 \\ -6 & -10 & 3 & 40 \end{pmatrix} \begin{pmatrix} 2 \\ 1 \\ x_3 \\ x_4 \end{pmatrix} = \begin{pmatrix} b_1 \\ b_2 \\ -7 \\ 61 \end{pmatrix}$$

Evaluate the unknowns b_1, b_2, x_3 and x_4.

The partitioning to be employed is obvious, giving

$$\begin{pmatrix} 40 & 3 \\ 3 & 10 \end{pmatrix} \begin{pmatrix} 2 \\ 1 \end{pmatrix} - \begin{pmatrix} 10 & 6 \\ 3 & 10 \end{pmatrix} \begin{pmatrix} x_3 \\ x_4 \end{pmatrix} = \begin{pmatrix} b_1 \\ b_2 \end{pmatrix}$$

and
$$-\begin{pmatrix} 10 & 3 \\ 6 & 10 \end{pmatrix} \begin{pmatrix} 2 \\ 1 \end{pmatrix} + \begin{pmatrix} 10 & 3 \\ 3 & 40 \end{pmatrix} \begin{pmatrix} x_3 \\ x_4 \end{pmatrix} = \begin{pmatrix} -7 \\ 61 \end{pmatrix}$$

From the second equation,

$$\begin{pmatrix} 10 & 3 \\ 3 & 40 \end{pmatrix} \begin{pmatrix} x_3 \\ x_4 \end{pmatrix} = \begin{pmatrix} -7 \\ 61 \end{pmatrix} + \begin{pmatrix} 10 & 3 \\ 6 & 10 \end{pmatrix} \begin{pmatrix} 2 \\ 1 \end{pmatrix} = \begin{pmatrix} 16 \\ 83 \end{pmatrix}$$

or
$$\begin{pmatrix} x_3 \\ x_4 \end{pmatrix} = \begin{pmatrix} 10 & 3 \\ 3 & 40 \end{pmatrix}^{-1} \begin{pmatrix} 16 \\ 83 \end{pmatrix} = \frac{1}{391} \begin{pmatrix} 40 & -3 \\ -3 & 10 \end{pmatrix} \begin{pmatrix} 16 \\ 83 \end{pmatrix}$$

$$= \frac{1}{391} \begin{pmatrix} 391 \\ 782 \end{pmatrix} = \begin{pmatrix} 1 \\ 2 \end{pmatrix}$$

Substituting into the first equation gives

$$\begin{pmatrix} b_1 \\ b_2 \end{pmatrix} = \begin{pmatrix} 40 & 3 \\ 3 & 10 \end{pmatrix} \begin{pmatrix} 2 \\ 1 \end{pmatrix} - \begin{pmatrix} 10 & 6 \\ 3 & 10 \end{pmatrix} \begin{pmatrix} 1 \\ 2 \end{pmatrix} = \begin{pmatrix} 83 \\ 16 \end{pmatrix} - \begin{pmatrix} 22 \\ 23 \end{pmatrix} = \begin{pmatrix} 61 \\ 7 \end{pmatrix}$$

Hence the unknowns are given by

$$\begin{pmatrix} b_1 \\ b_2 \end{pmatrix} = \begin{pmatrix} 61 \\ -7 \end{pmatrix} \text{ and } \begin{pmatrix} x_3 \\ x_4 \end{pmatrix} = \begin{pmatrix} 1 \\ 2 \end{pmatrix}$$

EXERCISES 2

1. Find the inverse of the following matrices, all of order 2, if they exist.

$$\begin{pmatrix} 1 & 2 \\ 3 & 7 \end{pmatrix}, \begin{pmatrix} 1 & 2 \\ 2 & 4 \end{pmatrix}, \begin{pmatrix} 3 & 5 \\ 7 & 9 \end{pmatrix} \text{ and } \begin{pmatrix} 3 & 2 \\ 1 & 8 \end{pmatrix}$$

2. Find the inverse of the following two matrices of order 3.

$$\begin{pmatrix} 2 & 2 & 1 \\ -2 & 1 & 2 \\ 1 & -2 & 2 \end{pmatrix} \text{ and } \begin{pmatrix} 1 & 2 & 1 \\ 1 & 2 & 0 \\ 1 & 4 & 2 \end{pmatrix}$$

3. Using the Gauss–Jordan elimination method, calculate the inverse of the following two matrices.

$$\begin{pmatrix} 1 & 1 & 1 & 1 \\ 1 & 2 & 3 & 4 \\ 1 & 3 & 6 & 10 \\ 1 & 4 & 10 & 20 \end{pmatrix} \text{ and } \begin{pmatrix} 1 & 2 & 3 & 4 \\ 1 & 3 & 6 & 2 \\ 1 & 3 & 7 & 8 \\ 1 & 3 & 2 & 9 \end{pmatrix}$$

4. An approximate inverse to the matrix $\mathbf{A}$, where

$$\mathbf{A} = \begin{pmatrix} 1 & 1 & 1 \\ 1 & 2 & 3 \\ 1 & 3 & 6 \end{pmatrix}, \text{ is given by } \mathbf{B} = \begin{pmatrix} 2{\cdot}9 & -3{\cdot}1 & 1 \\ -3{\cdot}1 & 5{\cdot}1 & -2{\cdot}1 \\ 1 & -1{\cdot}9 & 1 \end{pmatrix}$$

Find a matrix which is an improvement on this as an approximate to to $\mathbf{A}^{-1}$.

5. By suitably partitioning or otherwise, find the inverse of two matrices

$$\begin{pmatrix} 1 & 2 & 0 & 0 \\ 3 & 5 & 0 & 0 \\ 3 & 4 & 3 & 4 \\ 6 & 7 & 2 & 3 \end{pmatrix} \text{ and } \begin{pmatrix} 1 & 1 & 1 & 1 \\ 1 & 2 & 1 & 1 \\ 0 & 0 & 2 & 1 \\ 0 & 0 & 5 & 3 \end{pmatrix}$$

6. A set of linear equations has the form

$$\begin{pmatrix} 200 & 10 & -100 & -20 \\ 10 & 100 & -10 & -100 \\ -100 & -10 & 100 & 10 \\ -20 & -100 & 10 & 200 \end{pmatrix} \begin{pmatrix} 0{\cdot}1 \\ 0{\cdot}3 \\ x_3 \\ x_4 \end{pmatrix} = \begin{pmatrix} b_1 \\ b_2 \\ 11 \\ 64 \end{pmatrix}$$

Calculate the unknowns b_1, b_2, x_3 and x_4.

3

LINEAR EQUATIONS

3.1. Introduction

Throughout this chapter we will be concerned with solving sets of linear equations of the general form

$$\mathbf{A}\,\mathbf{X} = \mathbf{B} \tag{3.1}$$

where $\mathbf{A}$ is a square matrix and both $\mathbf{X}$ and $\mathbf{B}$ are vectors. Two distinct cases exist, one when $\mathbf{B}$ is the null vector and one when it is not. If $\mathbf{B} = \mathbf{O}$ the set of equations are said to be HOMOGENEOUS while the set is NON-HOMOGENEOUS if $\mathbf{B} \neq \mathbf{O}$. Non-homogeneous equations have already been mentioned briefly in Chapter 2; we shall return to discuss them further in the latter part of the chapter, turning first to investigate homogeneous equations.

3.2. Homogeneous equations

A set of homogeneous equations can always be written in the form

$$\mathbf{A}\,\mathbf{X} = \mathbf{O} \tag{3.2}$$

Clearly a solution always exists, namely,

$$\mathbf{X} = \mathbf{O}$$

Such a solution is called a TRIVIAL solution. The problem is to find other, NON-TRIVIAL, solutions if they exist. Non-trivial solutions need not exist; for example, the set of equations

$$x + y = 0$$
$$x + 2y = 0$$

has only the solution $x = 0$ and $y = 0$.

If a non-trivial solution does exist, then it is not unique. Let $\mathbf{X}_1$ be such a solution so that

$$\mathbf{A}\,\mathbf{X}_1 = \mathbf{O}$$

Multiplying this equation by any non-zero scalar λ gives

$$\lambda\,\mathbf{A}\,\mathbf{X}_1 = \mathbf{A}(\lambda\,\mathbf{X}_1) = \mathbf{O}$$

which shows that $\lambda\mathbf{X}_1$ is also a solution of the original equation.

The existence or otherwise of non-trivial solutions to equation (3.2) clearly depends only on the matrix $\mathbf{A}$ involved. Consider first the case when $\mathbf{A}$ is a non-singular matrix. For this case, the inverse of $\mathbf{A}$, $\mathbf{A}^{-1}$, exists. Hence, multiplying equation (3.2) by $\mathbf{A}^{-1}$ gives

$$\mathbf{A}^{-1}\,\mathbf{A}\,\mathbf{X} = \mathbf{I}\,\mathbf{X} = \mathbf{X} = \mathbf{O}$$

or

$$\mathbf{X} = \mathbf{O}$$

It has therefore been demonstrated that non-trivial solutions cannot exist if $\mathbf{A}$ is a non-singular matrix and so we need only consider the existence of non-trivial solutions when $\mathbf{A}$ is a singular matrix. It is possible to prove that if $\mathbf{A}$ is a singular matrix so that $|\mathbf{A}| = \mathbf{O}$, then non-trivial solutions will always exist. The proof is by induction, the theorem obviously being true for one equation in one unknown. It is, however, a very long proof and has been omitted. Therefore, we take it as having been proved that non-trivial solutions exist if, and only if, $\mathbf{A}$ is a singular matrix.

The reason for the validity of this theorem is obvious. If $\mathbf{A}$ is a singular matrix then, in some sense, at least one of its rows is dependent on the other rows so that at least one of the original equations carries no information. Removal of this equation (assuming it can be located) leaves the amount of information available unaltered and, after an arbitrary choice of value for one of the unknowns has been made, the remaining equations form a non-homogeneous set which can be solved. However, in view of the arbitrary choice of value for one unknown, this solution will be non-unique.

The above argument therefore suggests that we should define this concept of linear dependence of vectors (i.e. of rows of matrices).

3.3. Linear dependence

If $\mathbf{X}_1$, $\mathbf{X}_2$, ..., $\mathbf{X}_3$ are m vectors, all of the same order, they are said to be LINEARLY DEPENDENT if, and only if, scalars c_1, c_2, ..., c_m, not all being zero, can be found such that

$$c_1\mathbf{X}_1 + c_2\mathbf{X}_2 + \dots + c_m\mathbf{X}_m = \mathbf{0} \tag{3.3}$$

The set of vectors would be LINEARLY INDEPENDENT if no such scalars can be found.

Consider now the set of n vectors $\mathbf{X}_1$, $\mathbf{X}_2$, ..., $\mathbf{X}_n$, all of order n. This set will be linearly dependent if, and only if, scalars c_1, ..., c_n can be found such that

$$c_1\mathbf{X}_1 + c_2\mathbf{X}_2 \dots c_n\mathbf{X}_n = \mathbf{0} \tag{3.4}$$

But equation (3.4) can be rewritten as

$$\mathbf{A}\,\mathbf{C} = \mathbf{0} \tag{3.5}$$

where $\mathbf{A}$ is the square matrix whose columns are the vectors $\mathbf{X}_1$ ... $\mathbf{X}_n$ and $\mathbf{C}$ is the vector where elements are c_1 ... c_n. Equations (3.5) are however a homogeneous set of equations and will have a non-trivial solution, that is non-zero values for $\mathbf{C}$, only if

$$|\mathbf{A}| = 0$$

Thus, the vectors $\mathbf{X}_1$ ... $\mathbf{X}_n$ form a dependent set if the value of the determinant with the vectors as its columns is zero. Otherwise, the vectors are linearly independent.

Let $\mathbf{X}_1$, ..., $\mathbf{X}_n$ be n independent vectors of order n. Then, if $\mathbf{A}$ is the matrix generated with $\mathbf{X}_1$, ..., $\mathbf{X}_n$ as its columns, $\mathbf{A}$ is a non-singular matrix and $\mathbf{A}^{-1}$ exists. Now let $\mathbf{Y}$ be any other vector also of order n. As $\mathbf{A}^{-1}$ exists, $\mathbf{A}^{-1}\,\mathbf{Y}$ will also exist and in fact

$$\mathbf{A}^{-1}\mathbf{Y} = \mathbf{C} \tag{3.6}$$

where $\mathbf{C}$ is some vector of order n. Equation (3.6) gives

$$\mathbf{Y} = \mathbf{A}\,\mathbf{C}$$

which means that

$$\mathbf{Y} = c_1\mathbf{X}_1 + c_2\mathbf{X}_2 \ldots + c_n\mathbf{X}_n \tag{3.7}$$

or $\mathbf{Y}$ is dependent on the vector $\mathbf{X}_1, \ldots, \mathbf{X}_n$. It has therefore been shown that only n vectors of order n can be linearly independent.

This is of course a standard result in vector algebra and means that in three dimensions any vector can be expressed in terms of three independent vectors (usually the coordinate axes).

Example 3.1: $\mathbf{X}_1$, $\mathbf{X}_2$ and $\mathbf{X}_3$ are the three vectors given by

$$\mathbf{X}_1 = \begin{pmatrix} 2 \\ -1 \\ 2 \end{pmatrix}, \mathbf{X}_2 = \begin{pmatrix} -1 \\ 2 \\ 2 \end{pmatrix} \text{ and } \mathbf{X}_3 = \begin{pmatrix} 2 \\ 2 \\ -1 \end{pmatrix}$$

Verify that they form a linearly independent set. Express

$$\mathbf{Y} = \begin{pmatrix} 9 \\ 18 \\ 81 \end{pmatrix} \text{ in terms of } \mathbf{X}_1, \mathbf{X}_2 \text{ and } \mathbf{X}_3.$$

$$\mathbf{A} = \begin{pmatrix} 2 & -1 & 2 \\ -1 & 2 & 2 \\ 2 & 2 & -1 \end{pmatrix}, \quad |\mathbf{A}| = \begin{vmatrix} 2 & -1 & 2 \\ -1 & 2 & 2 \\ 2 & 2 & -1 \end{vmatrix} = -27 \neq 0$$

Therefore $\mathbf{X}_1$, $\mathbf{X}_2$ and $\mathbf{X}_3$ form a linearly independent set of vectors.

$$\mathbf{A}^{-1} = \begin{pmatrix} 2 & -1 & 2 \\ -1 & 2 & 2 \\ 2 & 2 & -1 \end{pmatrix}^{-1} = \frac{1}{9}\begin{pmatrix} 2 & -1 & 2 \\ -1 & 2 & 2 \\ 2 & 2 & -1 \end{pmatrix}$$

$$\mathbf{A}^{-1}\mathbf{Y} = \frac{1}{9}\begin{pmatrix} 2 & -1 & 2 \\ -1 & 2 & 2 \\ 2 & 2 & -1 \end{pmatrix}\begin{pmatrix} 9 \\ 18 \\ 81 \end{pmatrix} = \begin{pmatrix} 18 \\ 21 \\ -3 \end{pmatrix}$$

Hence

$$\mathbf{Y} = \mathbf{A}\begin{pmatrix} 18 \\ 21 \\ -3 \end{pmatrix} = (\mathbf{X}_1\mathbf{X}_2\mathbf{X}_3)\begin{pmatrix} 18 \\ 21 \\ -3 \end{pmatrix} = 18\mathbf{X}_1 + 21\mathbf{X}_2 - 3\mathbf{X}_3$$

$$\text{or}\quad \begin{pmatrix} 9 \\ 18 \\ 81 \end{pmatrix} = 18\begin{pmatrix} 2 \\ -1 \\ 2 \end{pmatrix} + 21\begin{pmatrix} -1 \\ 2 \\ 2 \end{pmatrix} - 3\begin{pmatrix} 2 \\ 2 \\ -1 \end{pmatrix}$$

3.4. Solution of homogeneous equations and rank

The work in section 3.2 allows us to determine whether a set of homogeneous equations has non-trivial solutions or not. It does not give a method for determining such solutions when they exist. The simplest way to obtain a solution is by systematic elimination, that is, the first equation is used to eliminate x_1 from all the remaining equations. The second equation is then used to eliminate x_2 from the remaining equations and so on. As the determinant of the coefficients of the equations is zero a stage will be reached where no further elimination is possible without generating $0 = 0$. When this occurs all the unknown in this equation except one can be assigned arbitrary values; the remaining unknowns of the problem can then be evaluated by substitution.

The RANK of a set of equations is the number of independent equations present. It is always less than or equal to the order of the equations. The rank is also equal to the order of the equations minus the number of unknowns that could be arbitrarily assigned.

Example 3.2: Solve the homogeneous set of equations

$$\begin{aligned} x_1 + 2x_2 + x_3 &= 0 \\ 2x_1 - x_2 + 3x_3 &= 0 \\ 3x_1 + x_2 + 4x_3 &= 0 \end{aligned}$$

Using the first equation to eliminate x_1 from the other two equations we obtain

$$-5x_2 + x_3 = 0$$

and

$$-5x_2 + x_3 = 0$$

No further elimination is possible and so the rank of this set is 2, the independent equations being the first and either of the second and third. Arbitrarily selecting $x_2 = a$, we have

$$x_3 = 5a$$

Substituting into the first equation gives

$$x_1 = -7a$$

The solution therefore is $\begin{pmatrix} x_1 \\ x_2 \\ x_3 \end{pmatrix} = a\begin{pmatrix} -7 \\ 1 \\ 5 \end{pmatrix}$, which illustrates the non-uniqueness of non-trivial solutions, a being any scalar. Since only one arbitrary choice was made, this also shows that the rank is 2, being the order (3) minus the number of arbitrary selections (1).

3.5. Non-homogeneous equations

For the remainder of this chapter we shall deal with non-homogeneous equations, that is equations of the type

$$\mathbf{A}\,\mathbf{X} = \mathbf{B} \tag{3.8}$$

where $\mathbf{B}$ is a non-zero vector, emphasizing methods for solution that are most useful when the calculations are to be carried out on a computer. The most important physical cases arise when $\mathbf{A}$ is a non-singular matrix. However, solutions may also be possible under certain circumstances if $\mathbf{A}$ is a singular matrix.

If $\mathbf{A}$ is a singular matrix, the rank of $\mathbf{A}$ is less than the order o of $\mathbf{A}$ so that at least one of its rows is dependent on the other rows. Solution is possible provided the corresponding elements of $\mathbf{B}$ are related in a similar way to the relationship between the dependent rows of $\mathbf{A}$. Thus, if after a number of eliminations, the left-hand side of the equation becomes zero then the right-hand side becomes zero at the same time. Solution in this case is similar to the solution for a homogeneous set, that is, in the last equation remaining before elimination generated a zero on both sides of the equality sign, all the unknowns except one are assigned arbitrary values. Substitution then provides the full solution.

Example 3.3: Solve the following equations:

$$\begin{aligned} x + y + z + w &= 4 \\ 2x + y + 2z + w &= 6 \\ 3x + 2y + z + 2w &= 8 \\ 2x + 2y + 0z + 2w &= 6 \end{aligned}$$

Using the first equation to eliminate x from the other three equations gives

$$\begin{aligned} y + w &= 2 \\ y + 2z + w &= 4 \\ 2z &= 2 \end{aligned}$$

Thus, $z = 1$, and the other two equations become

$$\begin{aligned} y + w &= 2 \\ y + w &= 2 \end{aligned}$$

Further elimination gives $0 = 0$ and so we arbitrarily assign $w = a$, then $y = 2 - a$. Substitution into the first equation gives $x = 1$ and the solution is

$$\begin{pmatrix} x \\ y \\ z \\ w \end{pmatrix} = \begin{pmatrix} 1 \\ 2-a \\ 1 \\ a \end{pmatrix}$$

3.6. Gaussian elimination

The most satisfactory way of solving equations of the form

$$\mathbf{A\,X} = \mathbf{B}$$

when $\mathbf{A}$ is non-singular, $\mathbf{B}$ is non-zero and the number of equations is large, is by the Gaussian elimination method. this is in fact the method from which the Gauss–Jordan method for calculating the inverse of a matrix was derived. The basis of the method is that of systematic elimination as in the Gauss–Jordan method, using row 1 to eliminate all elements in column 1 except the one in row 1. The elimination is repeated using row 2 in column 2 and so on. The difference now is that on solving

$$\mathbf{A\,X} = \mathbf{B}$$

with its solution

$$\mathbf{I\,X} = \mathbf{A}^{-1}\,\mathbf{B} = \mathbf{C} \tag{3.9}$$

the subsidiary step of obtaining $\mathbf{A}^{-1}$ is not required, only the final vector $\mathbf{A}^{-1}\,\mathbf{B} = \mathbf{C}$. Hence, the original vector $\mathbf{B}$ is set up on the right-hand side rather than the unit matrix as was the case in the method of Gauss–Jordan. The columns may now be interchanged as this only changes the order of the unknowns. They may not be multiplied or added together under any circumstances.

Example 3.4: Solve the system of equations

$$\begin{aligned} x_1 + x_2 + x_3 &= 6 \\ x_1 + 2x_2 + 2x_3 &= 11 \\ x_1 + 3x_2 + 2x_3 &= 13 \end{aligned}$$

The problem may be set up in the same manner as the problem was set up in the Gauss–Jordan method, namely, as

$$\begin{array}{ccc|c} 1 & 1 & 1 & 6 \\ 1 & 2 & 2 & 11 \\ 1 & 3 & 2 & 13 \\ \hline x_1 & x_2 & x_3 & \end{array}$$

Now, however, the unknowns which correspond to the columns have been noted as column interchanges are allowed. Using the same notation as for the Gauss–Jordan method we have

$$R_2 \rightarrow R_2 - R_1,\ R_3 \rightarrow R_3 - R_1$$

$$\begin{array}{ccc|c} 1 & 1 & 1 & 6 \\ 0 & 1 & 1 & 5 \\ 0 & 2 & 1 & 7 \\ \hline x_1 & x_2 & x_3 & \end{array}$$

Denoting in a similar way operations on columns, we have

$$C_2 \rightarrow C_3,\ C_3 \rightarrow C_2 \qquad\qquad R_3 \rightarrow R_3 - R_2,\ R_1 \rightarrow R_1 - R_2$$

$$\begin{array}{ccc|c} 1 & 1 & 1 & 6 \\ 0 & 1 & 1 & 5 \\ 0 & 1 & 2 & 7 \\ \hline x_1 & x_3 & x_2 & \end{array} \qquad\qquad \begin{array}{ccc|c} 1 & 0 & 0 & 1 \\ 0 & 1 & 1 & 5 \\ 0 & 0 & 1 & 2 \\ \hline x_1 & x_3 & x_2 & \end{array}$$

$$R_2 \rightarrow R_2 - R_3$$

$$\begin{array}{ccc|c} 1 & 0 & 0 & 1 \\ 0 & 1 & 0 & 3 \\ 0 & 0 & 1 & 2 \\ \hline x_1 & x_3 & x_2 & \end{array}$$

The solution is therefore given by

$$\begin{pmatrix} x_1 \\ x_2 \\ x_3 \end{pmatrix} = \begin{pmatrix} 1 \\ 2 \\ 3 \end{pmatrix}$$

3.7. Back substitution

It is not necessary to reduce the left-hand array to a diagonal form. If it has been reduced to an upper triangular form, the last row will be sufficient to determine one unknown. Substitution of this value into the next but last row allows the determination of another unknown. Further BACK SUBSTITUTION will evaluate all the unknowns. The amount of manipulation necessary to evaluate the unknowns by back substitution is in fact the same as is necessary to eliminate the coefficients so that no real advantage is gained by using back substitution when the method of Gaussian elimination is being used. However, should the matrix of coefficients be an upper triangular matrix or easily reducable to an upper triangular matrix, then back substitution is advantageous.

3.8. Pivoting

As in the Gauss–Jordan method, round-off errors can cause a considerable inaccuracy if division by small numbers takes place. It is therefore advisable to pivot whenever the Gaussian elimination method is being employed. Now, however, columns are interchangeable as well as rows and so the largest element can be moved to the top left-hand corner of the array under discussion and not just onto the leading diagonal as was the case when the Gauss–Jordan method was discussed. As before once a row has been used it is disregarded in the search for the largest element.

Example 3.5: Solve the following set of equations

$$\begin{aligned} 2x_1 + x_2 + x_3 &= 8 \\ x_1 + 2x_2 + 2x_3 &= 17 \\ x_1 + 2x_2 + 3x_3 &= 17 \end{aligned}$$

As in previous examples we use R_i to denote row i and C_i to denote column i. All the elements are again integers and we will preserve them as integers by again considering operations of the form $aR_i - bR_j$ rather than of the form $R_i - (b/a)R_j$ which is the operation that actually occurs in a computer evaluation.

original

x_1	x_2	x_3	
2	1	1	8
1	2	2	13
1	2	3	17

$R_1 \to R_3,\ R_3 \to R_1$

x_1	x_2	x_3	
1	2	3	17
1	2	2	13
2	1	1	8

This first move, and the following move, is carried out in order to place the largest element (3 in row 3 column 3) in the top left-hand corner, followed by an elimination in column 1.

$C_1 \to C_3,\ C_3 \to C_1$

x_3	x_2	x_1	
3	2	1	17
2	2	1	13
1	1	2	8

$R_2 \to 3R_2 - 2R_1,\ R_3 \to 3R_3 - R_1$

x_3	x_2	x_1	
3	2	1	17
0	2	1	5
0	1	5	7

In the array now available (rows 2 and 3) the largest element is 5 and the following two operations moves it to the highest left-hand corner available

$R_2 \to R_3,\ R_3 \to R_2$

x_3	x_2	x_1	
3	2	1	17
0	1	5	7
0	2	1	5

$C_3 \to C_2,\ C_2 \to C_3$

x_3	x_1	x_2	
3	1	2	17
0	5	1	7
0	1	2	5

$R_3 \to 5R_3 - R_2$

x_3	x_1	x_2	
3	1	2	17
0	5	1	7
0	0	9	18

From the last row we have $x_2 = 2$. By back substitution $x_1 = 1$ and $x_3 = 4$ and so the solution is

$$\begin{pmatrix} x_1 \\ x_2 \\ x_3 \end{pmatrix} = \begin{pmatrix} 1 \\ 2 \\ 4 \end{pmatrix}$$

3.9. Decomposition or L U method

This method of solving equations is very elegant and efficient especially when the matrix of coefficients is a banded matrix. It was shown in Chapter 1 that any square matrix $\mathbf{A}$ can be represented as the product of a unit lower triangular matrix and an upper triangular matrix, that is

$$\mathbf{A} = \mathbf{L}\,\mathbf{U} \tag{3.10}$$

Any vector $\mathbf{B}$ can be written as

$$\mathbf{B} = \mathbf{L}\,\mathbf{C} \tag{3.11}$$

where $\mathbf{C}$ is a vector and $\mathbf{L}$ is a given unit lower triangular matrix. The elements $\mathbf{C}$ can be determined by starting with the first row where

$$b_1 = l_{11}\, c_1$$

thus determining c_1. The second row gives

$$b_2 = l_{21}\, c_1 + l_{22}\, c_2$$

which now determines c_2. Thus, row i gives

$$b_i = l_{i1} c_1 + l_{i2} c_2 + \ldots\ l_{ii} c_i$$

which determines c_i as $c_1 \ldots c_{i-1}$ have already been evaluated. Hence it has been shown that the unknown vector $\mathbf{C}$ of equation (3.11) can be determined when $\mathbf{L}$ and $\mathbf{B}$ are given. An equation of the form

$$\mathbf{A}\,\mathbf{X} = \mathbf{B} \tag{3.12}$$

can therefore be written as

$$\mathbf{L}\,\mathbf{U}\,\mathbf{X} = \mathbf{L}\,\mathbf{C}$$

or

$$\mathbf{U}\,\mathbf{X} = \mathbf{C} \tag{3.13}$$

But this equation can be solved directly by back substitution and so the solution is obtained without elimination. This method of solution is particularly useful when the matrix of coefficients is a banded matrix as the upper and lower triangular matrices will not have elements outside the band either. Both the evaluation of the elements of **C** and the solution by back substitution then become very much simpler.

Example 3.6: Solve the system of linear equations

$$\begin{aligned} 2x_1 + 2x_2 + 3x_3 &= 15 \\ 6x_1 + 9x_2 + 13x_3 &= 63 \\ 10x_1 + 13x_2 + 24x_3 &= 108 \end{aligned}$$

The system is equivalent to

$$\begin{pmatrix} 2 & 2 & 3 \\ 6 & 9 & 13 \\ 10 & 13 & 24 \end{pmatrix} \begin{pmatrix} x_1 \\ x_2 \\ x_3 \end{pmatrix} = \begin{pmatrix} 15 \\ 63 \\ 108 \end{pmatrix}$$

By the methods described in section 1.8 it is easy to express the matrix as a product of a unit lower triangular matrix and an upper triangular matrix as

$$\begin{pmatrix} 2 & 2 & 3 \\ 6 & 9 & 13 \\ 10 & 13 & 24 \end{pmatrix} = \begin{pmatrix} 1 & 0 & 0 \\ 3 & 1 & 0 \\ 5 & 1 & 1 \end{pmatrix} \begin{pmatrix} 2 & 2 & 3 \\ 0 & 3 & 4 \\ 0 & 0 & 5 \end{pmatrix}$$

while by the methods just described

$$\begin{pmatrix} 15 \\ 63 \\ 108 \end{pmatrix} = \begin{pmatrix} 1 & 0 & 0 \\ 3 & 1 & 0 \\ 5 & 1 & 1 \end{pmatrix} \begin{pmatrix} 15 \\ 18 \\ 15 \end{pmatrix}$$

The system of equations is therefore equivalent to

$$\begin{pmatrix} 2 & 2 & 3 \\ 0 & 3 & 4 \\ 0 & 0 & 5 \end{pmatrix} \begin{pmatrix} x_1 \\ x_2 \\ x_3 \end{pmatrix} = \begin{pmatrix} 15 \\ 18 \\ 15 \end{pmatrix}$$

From the last row, $x_3 = 3$ and by back substitution $x_2 = 2$ and $x_1 = 1$ and so the solution is

$$\begin{pmatrix} x_1 \\ x_2 \\ x_3 \end{pmatrix} = \begin{pmatrix} 1 \\ 2 \\ 3 \end{pmatrix}$$

Example 3.7: Solve the system of linear equations

$$\begin{aligned} x_1 + 2x_2 + 0x_3 + 0x_4 &= 5 \\ 2x_1 + 6x_2 + 2x_3 + 0x_4 &= 20 \\ 0x_1 + 8x_2 + x_3 + 2x_4 &= 27 \\ 0x_1 + 0x_2 + 7x_3 + x_4 &= 25 \end{aligned}$$

The system is equivalent to

$$\begin{pmatrix} 1 & 2 & 0 & 0 \\ 2 & 6 & 2 & 0 \\ 0 & 8 & 1 & 2 \\ 0 & 0 & 7 & 1 \end{pmatrix} \begin{pmatrix} x_1 \\ x_2 \\ x_3 \\ x_4 \end{pmatrix} = \begin{pmatrix} 5 \\ 20 \\ 27 \\ 25 \end{pmatrix}$$

By the methods of section 1.8 we have

$$\begin{pmatrix} 1 & 2 & 0 & 0 \\ 2 & 6 & 2 & 0 \\ 0 & 8 & 1 & 2 \\ 0 & 0 & 7 & 1 \end{pmatrix} = \begin{pmatrix} 1 & 0 & 0 & 0 \\ 2 & 1 & 0 & 0 \\ 0 & 4 & 1 & 0 \\ 0 & 0 & -1 & 1 \end{pmatrix} \begin{pmatrix} 1 & 2 & 0 & 0 \\ 0 & 2 & 2 & 0 \\ 0 & 0 & -7 & 2 \\ 0 & 0 & 0 & 3 \end{pmatrix}$$

while

$$\begin{pmatrix} 5 \\ 20 \\ 27 \\ 25 \end{pmatrix} = \begin{pmatrix} 1 & 0 & 0 & 0 \\ 2 & 1 & 0 & 0 \\ 0 & 4 & 1 & 0 \\ 0 & 0 & -1 & 1 \end{pmatrix} \begin{pmatrix} 5 \\ 10 \\ -13 \\ 12 \end{pmatrix}$$

The system of equations is therefore equivalent to

$$\begin{pmatrix} 1 & 2 & 0 & 0 \\ 0 & 2 & 2 & 0 \\ 0 & 0 & -7 & 2 \\ 0 & 0 & 0 & 3 \end{pmatrix} \begin{pmatrix} x_1 \\ x_2 \\ x_3 \\ x_4 \end{pmatrix} = \begin{pmatrix} 5 \\ 10 \\ -13 \\ 12 \end{pmatrix}$$

The last row gives $x_4 = 4$ and by back substitution, $x_3 = 3$, $x_2 = 2$ and $x_1 = 1$, the solution therefore is

$$\begin{pmatrix} x_1 \\ x_2 \\ x_3 \\ x_4 \end{pmatrix} = \begin{pmatrix} 1 \\ 2 \\ 3 \\ 4 \end{pmatrix}$$

3.10 Iterative methods for obtaining a solution

For some types of problems it is easier to use an iterative method to obtain a solution to a system of equations than it is to use direct methods of solution such as the Gaussian elimination method. In order to obtain a solution by any iterative method, one approximate solution is required and an improved solution is then obtained from this. The following method is very useful if the matrix, $\mathbf{A}$, is diagonally dominant (as the stiffness matrix in structure problems tends to be, for example). There are numerous other interative methods available but unfortunately space will not allow the inclusion of them all.

Let the equations that are to be solved be of the form

$$\mathbf{A}\,\mathbf{X} = \mathbf{B} \tag{3.14}$$

and let

$$\mathbf{X} = \mathbf{Y} + \mathbf{C} \tag{3.15}$$

where $\mathbf{Y}$ is the vector whose components are given by

$$y_i = b_i/a_{ii}$$

that is $\mathbf{Y}$ is the vector which would have been the exact solution if the matrix $\mathbf{A}$ had been a diagonal matrix. $\mathbf{C}$ is a correction to allow for the presence of the non-diagonal terms and it has to be determined.

Substitution into the original equation gives

$$\mathbf{A}\,(\mathbf{Y} + \mathbf{C}) = \mathbf{B}$$

or

$$\mathbf{A}\,\mathbf{C} = \mathbf{B} - \mathbf{A}\,\mathbf{Y} = \mathbf{B}' \tag{3.16}$$

where $\mathbf{B}'$ is a calculable vector. Equation (3.16) is exactly similar to equation (3.14) and so an estimate for $\mathbf{C}$ is given by

$$\mathbf{C} = \mathbf{Y}' + \mathbf{C}',$$

where $\mathbf{Y}'$ is the vector whose elements are given by

$$y_i' = b_i'/a_{ii}$$

and $\mathbf{C}'$ is another correction vector. After a number of repeats the solution to the original equation is given by

$$\mathbf{X} = \mathbf{Y} + \mathbf{Y}' + \dots$$

Care must be taken to ensure that this method is only used when the matrix is diagonally dominant otherwise the process will not converge.

Example 3.8: Obtain an approximate solution to the following set set of linear equations

$$\begin{aligned} 10x_1 + x_2 + x_3 &= 15 \\ x_1 + 10x_2 + x_3 &= 24 \\ x_1 + x_2 + 10x_3 &= 33 \end{aligned}$$

The equations can be written in the form

$$\begin{pmatrix} 10 & 1 & 1 \\ 1 & 10 & 1 \\ 1 & 1 & 10 \end{pmatrix} \begin{pmatrix} x_1 \\ x_2 \\ x_3 \end{pmatrix} = \begin{pmatrix} 15 \\ 24 \\ 33 \end{pmatrix}$$

and so we note that the matrix that has been generated is diagonally dominant. Let

$$\begin{pmatrix} x_1 \\ x_2 \\ x_3 \end{pmatrix} = \begin{pmatrix} 1{\cdot}5 \\ 2{\cdot}4 \\ 3{\cdot}3 \end{pmatrix} + \begin{pmatrix} c_1 \\ c_2 \\ c_3 \end{pmatrix}$$

and substitute into the original equation. This gives

$$\begin{pmatrix} 10 & 1 & 1 \\ 1 & 10 & 1 \\ 1 & 1 & 10 \end{pmatrix} \begin{pmatrix} c_1 \\ c_2 \\ c_3 \end{pmatrix} = \begin{pmatrix} 15 \\ 24 \\ 33 \end{pmatrix} - \begin{pmatrix} 10 & 1 & 1 \\ 1 & 10 & 1 \\ 1 & 1 & 10 \end{pmatrix} \begin{pmatrix} 1{\cdot}5 \\ 2{\cdot}4 \\ 3{\cdot}3 \end{pmatrix} = -\begin{pmatrix} 5{\cdot}7 \\ 4{\cdot}8 \\ 3{\cdot}9 \end{pmatrix}$$

Now let

$$\begin{pmatrix} c_1 \\ c_2 \\ c_3 \end{pmatrix} = -\begin{pmatrix} 0{\cdot}57 \\ 0{\cdot}48 \\ 0{\cdot}39 \end{pmatrix} + \begin{pmatrix} c_1' \\ c_2' \\ c_3' \end{pmatrix}$$

and substitute into the above equation, giving

$$\begin{pmatrix}10 & 1 & 1\\ 1 & 10 & 1\\ 1 & 1 & 10\end{pmatrix}\begin{pmatrix}c'_1\\ c'_2\\ c'_3\end{pmatrix} = -\begin{pmatrix}5\cdot7\\ 4\cdot8\\ 3\cdot9\end{pmatrix} + \begin{pmatrix}10 & 1 & 1\\ 1 & 10 & 1\\ 1 & 1 & 10\end{pmatrix}\begin{pmatrix}0\cdot57\\ 0\cdot48\\ 0\cdot39\end{pmatrix} = \begin{pmatrix}0\cdot87\\ 0\cdot96\\ 1\cdot05\end{pmatrix}$$

Let

$$\begin{pmatrix}c'_1\\ c'_2\\ c'_3\end{pmatrix} = \begin{pmatrix}0\cdot087\\ 0\cdot096\\ 0\cdot105\end{pmatrix} + \begin{pmatrix}c''_1\\ c''_2\\ c''_3\end{pmatrix}$$

Hence, while a first approximation is $\begin{pmatrix}1\cdot5\\ 2\cdot4\\ 3\cdot3\end{pmatrix}$

a better approximation is given by $\begin{pmatrix}1\cdot5\\ 2\cdot4\\ 3\cdot3\end{pmatrix} - \begin{pmatrix}0\cdot57\\ 0\cdot48\\ 0\cdot39\end{pmatrix} = \begin{pmatrix}0\cdot93\\ 1\cdot92\\ 2\cdot91\end{pmatrix}$

while an even better approximation is

$$\begin{pmatrix}0\cdot93\\ 1\cdot92\\ 2\cdot91\end{pmatrix} + \begin{pmatrix}0\cdot087\\ 0\cdot096\\ 0\cdot105\end{pmatrix} = \begin{pmatrix}1\cdot017\\ 2\cdot016\\ 3\cdot015\end{pmatrix}$$

Further substitution to evaluate $\mathbf{C}''$ would improve the accuracy of the approximation even further. These equations can of course be solved by any of the other methods in this chapter and the correct solution is

$$\begin{pmatrix}x_1\\ x_2\\ x_3\end{pmatrix} = \begin{pmatrix}1\\ 2\\ 3\end{pmatrix}$$

3.11. The Gauss–Seidel method

This method is in many ways a modification of the above method but is far more efficient.

Consider again the set of equations

$$\mathbf{A}\,\mathbf{X} = \mathbf{B} \tag{3.17}$$

The value of x_1 would be given by

$$x_1 = \frac{b_1 - (a_{12}\ x_2 + \dots + a_{ln}\ x_n)}{a_{11}} \qquad (3.18)$$

if all the other values $x_2, \dots, x_n$ were known, with similar formulae for all the other unknowns, so that in general

$$x_i = \frac{1}{a_{ii}} [b_i - \sum_{j \neq i} a_{ij}\ x_j] \qquad (3.19)$$

A possible iterative process is therefore to select an arbitrary initial value for $\mathbf{X}$, $\mathbf{X}^\circ$ say; then, on using equation (3.19), a new approximate solution of $\mathbf{X}$, $\mathbf{X}^1$, can be found. It can be shown that, after a sufficient number of repeats, this iterative method also converges provided the original matrix $\mathbf{A}$ is diagonal dominant. It is clear that this method is not as efficient as it might be, for in the above calculation for the first iteration of x_i, the initial values of all the other unknowns were used, whereas in fact improved values for x_j for all $j < i$ are available. An improvement on the iteration of equation (3.19) is therefore given by

$$x_i^{k+1} = \frac{1}{a_{ii}} \left[b_i - \sum_{j=1}^{i-1} a_{ij}\ x_j^{k+1} - \sum_{j=i+1}^{n} a_{ij}\ x_j^k \right] \qquad (3.20)$$

where x_i^k denotes the kth iteration

Example 3.9: Find an approximate solution to the following set of linear equations

$$\begin{aligned} 6x_1 + 2x_2 - 3x_3 &= 5 \\ -x_1 + 8x_2 + 3x_3 &= -10 \\ x_1 + 4x_2 + 12x_3 &= 12 \end{aligned}$$

by using the Gauss–Seidel method.

Take as an arbitrary initial solution

$$\begin{pmatrix} x_1^\circ \\ x_2^\circ \\ x_3^\circ \end{pmatrix} = \begin{pmatrix} 1 \\ 1 \\ 1 \end{pmatrix}$$

Then, $x_1^1 = \frac{1}{6} [5 - 2 + 3] = 1$, on using the initial values of x_2° and x_3°. Also

$$x_2^1 = \frac{1}{8} [-10 + 1 - 3] = -3/2$$

on using the initial value of x_3, but the improved value of x_1. Similarly

$$x_3^1 = \frac{1}{12}[12 - 1 + 6] = \frac{17}{12}$$

on using the improved values of both x_1 and x_2. A second iteration gives

$$x_1^2 = \frac{1}{6}\left[5 + 3 + \frac{17}{4}\right] = \frac{49}{24}$$

$$x_2^2 = \frac{1}{8}\left[-10 + \frac{49}{24} - \frac{17}{4}\right] = -\frac{293}{192}$$

$$x_3^2 = \frac{1}{12}\left[12 - \frac{49}{24} + \frac{293}{48}\right] = \frac{257}{192}$$

A third iteration yields

$$x_1^3 = 2{\cdot}012,\ x_2^3 = -1{\cdot}500 \text{ and } x_3^3 = 1{\cdot}332$$

a reasonable approximation to the true solution

$$x_1 = 2,\ x_2 = -3/2 \text{ and } x_3 = 4/3$$

EXERCISES 3

1. Are the vectors

$$\begin{pmatrix}1\\1\\1\end{pmatrix}, \begin{pmatrix}1\\2\\2\end{pmatrix} \text{ and } \begin{pmatrix}1\\0\\1\end{pmatrix}$$

linearly independent? If they are, express both of the vectors

$$\begin{pmatrix}2\\7\\6\end{pmatrix} \text{ and } \begin{pmatrix}6\\5\\8\end{pmatrix}$$

in terms of them.

2. Solve the set of homogeneous equations

$$\begin{aligned} x_1 + x_2 + x_3 &= 0 \\ 3x_1 + x_2 - x_3 &= 0 \\ 5x_1 + 3x_2 + x_3 &= 0 \end{aligned}$$

What is the rank of this set?

3. Solve the set of non-homogeneous equations

$$\begin{aligned} x_1 + x_2 + x_3 &= 6 \\ x_1 + 2x_2 + 3x_3 &= 11 \\ 3x_1 + 4x_2 + 5x_3 &= 23 \end{aligned}$$

What is the rank of this set?

4. By the Gaussian elimination method solve each of the following sets of equations:

(a)
$$\begin{aligned} x_1 + 2x_2 + x_3 &= 4 \\ 2x_1 + x_2 + 3x_3 &= 6 \\ x_1 + 4x_2 - x_3 &= 4 \end{aligned}$$

(b)
$$\begin{aligned} 2x_1 + x_2 + 2x_3 &= 17 \\ x_1 + 2x_2 + 2x_3 &= 19 \\ 5x_1 + x_2 + x_3 &= 14 \end{aligned}$$

(c)
$$\begin{aligned} 3x_1 + 2x_2 + x_3 &= 13 \\ x_1 + 3x_2 + x_3 &= 8 \\ 2x_1 + 5x_2 + 2x_3 &= 15 \end{aligned}$$

5. By the **L U** method solve each of the following sets of equations

(a)
$$\begin{aligned} x_1 + 4x_2 + 5x_3 &= 14 \\ 2x_1 + 10x_2 + 16x_3 &= 38 \\ 3x_1 + 20x_2 + 42x_3 &= 85 \end{aligned}$$

(b)
$$\begin{aligned} 2x_1 + 2x_2 + 3x_3 &= 36 \\ 4x_1 + 8x_2 + 5x_3 &= 84 \\ 4x_1 + 10x_2 + 8x_3 &= 118 \end{aligned}$$

(c)
$$\begin{aligned} 7x_1 + x_2 + 3x_3 &= 21 \\ 21x_1 + 6x_2 + 11x_3 &= 77 \\ 35x_1 + 14x_2 + 25x_3 &= 163 \end{aligned}$$

6. By iteration, solve the following two sets of equations:

(a)
$$\begin{aligned} 10x_1 + x_2 + x_3 &= 12 \\ x_1 + 10x_2 + x_3 &= 12 \\ x_1 + x_2 + 10x_3 &= 12 \end{aligned}$$

(b)
$$\begin{aligned} 5x_1 + x_2 + x_3 &= 7 \\ x_1 + 5x_2 + x_3 &= 7 \\ x_1 + x_2 + 5x_3 &= 7 \end{aligned}$$

4

EIGENVALUES AND EIGENVECTORS

4.1. The eigenvalues of a matrix

There would be a considerable simplification in the algebra if, under certain circumstances, the product $\mathbf{A}\,\mathbf{X}$, where $\mathbf{A}$ is a square matrix and $\mathbf{X}$ a column vector, could be written as $\lambda\mathbf{X}$, where λ is a scalar. This has considerable physical significance, most of which will become apparent when the theory has been developed and the examples in Chapter 6 studied. One simple implication of having

$$\mathbf{A}\,\mathbf{X} = \lambda\mathbf{X} \tag{4.1}$$

is that when a vector $\mathbf{X}$ is transformed by the matrix $\mathbf{A}$, the resultant vector remains parallel to the original vector. Equation (4.1) may be rewritten as

$$(\mathbf{A} - \lambda\,\mathbf{I})\,\mathbf{X} = 0 \tag{4.2}$$

This is, however, equivalent to a set of homogeneous equations and a non-zero value for the vector $\mathbf{X}$, which corresponds to non-trivial solutions to the set of equations, can exist if, and only if,

$$|\mathbf{A} - \lambda\,\mathbf{I}| = 0$$

Writing this determinant out more fully gives

$$\begin{vmatrix} a_{11}-\lambda & a_{12} & \dots & a_{1n} \\ a_{21} & a_{22}-\lambda & \dots & a_{2n} \\ \dots & \dots & \dots & \dots \\ a_{n1} & a_{n2} & \dots & a_{nn}-\lambda \end{vmatrix} = 0 \tag{4.3}$$

Evaluation of this determinant will generate a polynomial of degree n in λ. This polynomial is called the CHARACTERISTIC EQUATION of the matrix and its n roots (which need not all be distinct) are called the EIGENVALUES (or latent roots, or characteristic roots). Non-zero vectors $\mathbf{X}$ will exist only when λ is one of the eigenvalues. It is clear from the rules for the expansion of a determinant that the terms in λ^n and λ^{n-1} will arise only from the product of all the terms in the leading diagonal while the term that is independent of λ will be $|\mathbf{A}|$. The characteristic equation therefore has the form

$$\lambda^n - (a_{11} + a_{22} \ldots + a_{nn})\lambda^{n-1} \ldots + (-1)^n |\mathbf{A}| = 0 \quad (4.4)$$

But, from elementary algebraic considerations, the sum of the roots of this equation, $\lambda_1 + \lambda_2 \ldots + \lambda_n$ say, is given by minus the ratio of the coefficients of λ^{n-1} to the coefficient of λ^n, that is

$$\lambda_1 + \lambda_2 \ldots + \lambda_n = a_{11} + a_{22} \ldots + a_{nn}$$

Hence, the sum of the eigenvalues of any matrix is equal to the sum of the elements in the leading diagonal of the matrix. This sum is called the TRACE of the matrix, or sometimes, the SPUR of the matrix, and will be denoted by Tr$\mathbf{A}$. From the same algebraic considerations, the product of the eigenvalues will be the determinant of the matrix, $|\mathbf{A}|$. Thus, if $\mathbf{A}$ is a non-singular matrix, none of its eigenvalues can be zero. These two results are of considerable importance when the eigenvalues have to be determined by numerical methods (this being the case for most physical situations) because, when the sum and the product of the eigenvalues are known, two of the n eigenvalues can be determined when the remaining n–2 eigenvalues are known.

4.2. Eigenvectors

The theorem on the existence of non-trivial solutions for a set of homogeneous equations that was used shows that corresponding to every value for λ that has been found there will exist a non-zero vector $\mathbf{X}$ such that

$$(\mathbf{A} - \lambda \mathbf{I})\mathbf{X} = 0, \text{ or } \mathbf{A}\,\mathbf{X} = \lambda\mathbf{X} \quad (4.5)$$

Such a column vector is called the COLUMN EIGENVECTOR (pole) corresponding to the given eigenvalue λ. The eigenvector is the non-trivial solution to a set of homogeneous equations and so is not unique, that is, a scalar multiple of any eigenvector will also be an eigenvector. In many cases it will not be necessary to solve the system of homogeneous equations by the methods described in Chapter 3, for a solution will be fairly obvious.

Now, as $|\mathbf{A} - \lambda\mathbf{I}| = 0$, $|(\mathbf{A} - \lambda\mathbf{I})^T|$ will also be zero and so will $|\mathbf{A}^T - \lambda\mathbf{I}|$. Hence, λ is also an eigenvalue of $\mathbf{A}^T$. This being the case, there exists some vector $\mathbf{Y}$ such that

$$(\mathbf{A}^T - \lambda\,\mathbf{I})\mathbf{Y} = \mathbf{0}$$

Transposing this equation gives

$$\mathbf{Y}^T(\mathbf{A}^T - \lambda\mathbf{I})^T = \mathbf{Y}^T(\mathbf{A} - \lambda\mathbf{I}) = 0$$

which shows that corresponding to the eigenvalue λ there also exists a row vector $\mathbf{Y}^T$, called the ROW EIGENVECTOR, such that

$$\mathbf{Y}^T\mathbf{A} = \lambda\mathbf{Y}^T$$

If $\mathbf{A}$ is a symmetric matrix, then the row eigenvector is the column eigenvector transposed, but not otherwise.

From now on the term eigenvector will be taken to mean the column eigenvector, specifying row or column only when both kinds are involved.

Example 4.1: Find the eigenvalues and corresponding row and column eigenvectors of the matrix $\mathbf{A}$, where

$$\mathbf{A} = \begin{pmatrix} 6 & -2 & 2 \\ -2 & 5 & 0 \\ 2 & 0 & 7 \end{pmatrix}$$

$$|\mathbf{A} - \lambda\mathbf{I}| = \begin{vmatrix} 6-\lambda & -2 & 2 \\ -2 & 5-\lambda & 0 \\ 2 & 0 & 7-\lambda \end{vmatrix} = 0$$

Evaluating this determinant gives the characteristic equation as

$$\lambda^3 - 18\lambda^2 + 99\lambda - 162 = 0$$

This cubic equation factorizes to give

$$(\lambda - 3)(\lambda - 6)(\lambda - 9) = 0$$

The three eigenvalues of this matrix are therefore 3, 6 and 9. We note that their sum is 18, the same as the trace of **A**.

$$\text{When } \lambda = 3, (\mathbf{A} - \lambda\mathbf{I})\,\mathbf{X} = \begin{pmatrix} 3 & -2 & 2 \\ -2 & 2 & 0 \\ 2 & 0 & 4 \end{pmatrix}\begin{pmatrix} x_1 \\ x_2 \\ x_3 \end{pmatrix} = 0$$

These equations give

$$\begin{aligned} 3x_1 - 2x_2 + 2x_3 &= 0 \\ x_1 &= x_2 \\ x_1 &= -2x_3 \end{aligned}$$

Hence, letting $x_3 = -\mu$, we obtain $x_1 = x_2 = 2\mu$. Note that this solution also satisfies $3x_1 - 2x_2 + 2x_3 = 0$. The column eigenvector is

$$\mu\begin{pmatrix} 2 \\ 2 \\ -1 \end{pmatrix}$$

where μ can take any value, thus demonstrating the non-uniqueness of the eigenvector. Since the matrix is symmetric, the row eigenvector is $\mu\,(2, 2, -1)$, and clearly

$$\mu(2, 2, -1)\begin{pmatrix} 3 & -2 & 2 \\ -2 & 2 & 0 \\ 2 & 0 & 4 \end{pmatrix} = (0\ 0\ 0)$$

$$\text{When } \lambda = 6, (\mathbf{A} - \lambda\mathbf{I})\,\mathbf{X} = \begin{pmatrix} 0 & -2 & 2 \\ -2 & -1 & 0 \\ 2 & 0 & 1 \end{pmatrix}\begin{pmatrix} x_1 \\ x_2 \\ x_3 \end{pmatrix} = \mathbf{0}$$

or $x_2 = x_3$, $2x_1 = -x_2$ and $2x_1 = -x_3$. Selecting $x_1 = -\mu$, when μ need not have the same value as in the case when $\lambda = 3$, the column eigenvector is given by

$$\mu\begin{pmatrix} -1 \\ 2 \\ 2 \end{pmatrix}$$

and again the row eigenvector is $\mu(-1, 2, 2)$.

$$\text{When } \lambda = 9, (\mathbf{A} - \lambda\mathbf{I})\,\mathbf{X} = \begin{pmatrix} -3 & -2 & 2 \\ -2 & -4 & 0 \\ 2 & 0 & -2 \end{pmatrix}\begin{pmatrix} x_1 \\ x_2 \\ x_3 \end{pmatrix} = \mathbf{0}$$

Hence, letting $x_2 = -\mu$, we have $x_1 = x_3 = 2\mu$ and the column eigenvector is

$$\mu \begin{pmatrix} 2 \\ -1 \\ 2 \end{pmatrix}$$

while the row eigenvector is $\mu(2 \; -1 \; 2)$.

4.3. Reduction to a diagonal matrix

If $\mathbf{A}$ is a matrix of order n for which there exists n linearly independent eigenvectors, then there exists a non-singular matrix $\mathbf{Q}$, also of order n, such that

$$\mathbf{Q}^{-1} \mathbf{A} \mathbf{Q} = \Lambda, \tag{4.6}$$

where Λ is a diagonal matrix.

To prove this statement, let $\lambda_1, \ldots, \lambda_n$ be the eigenvalues of $\mathbf{A}$ and let $\mathbf{X}_1, \ldots, \mathbf{X}_n$ be the n linearly independent eigenvectors Now, construct a matrix $\mathbf{Q}$ with the n eigenvectors as its columns. As the eigenvectors are linearly independent, $\mathbf{Q}$ will be a non-singular matrix.

Now, for every eigenvector $\mathbf{X}_i$,

$$\mathbf{A} \mathbf{X}_i = \lambda_i \mathbf{X}_i$$

and so $\mathbf{A} \mathbf{Q} = \mathbf{A}(\mathbf{X}_1 \ldots \mathbf{X}_n) = (\lambda_1 \mathbf{X}_1 \ldots \lambda_n \mathbf{X}_n) =$ $(\mathbf{X}_1 \ldots \mathbf{X}_n) \, \mathbf{diag}\,(\lambda) = \mathbf{Q} \, \mathbf{diag}\,(\lambda)$. But, as $\mathbf{Q}$ is non-singular, $\mathbf{Q}^{-1}$ will exist and so

$$\mathbf{Q}^{-1} \mathbf{A} \mathbf{Q} = \mathbf{diag}\,\lambda = \Lambda$$

thus proving statement (4.6)

4.4. Linear dependence of the eigenvectors

The validity of the above theorem depends on the existence of a set of linearly independent eigenvectors. We shall show that such a set always exists if all the eigenvalues of the matrix are distinct.

Let $\mathbf{X}_1, \mathbf{X}_2, \ldots, \mathbf{X}_n$ be the n eigenvectors corresponding to the distinct eigenvalues $\lambda_1, \lambda_2, \ldots, \lambda_n$. It is required to show that $\mathbf{X}_1, \mathbf{X}_2, \ldots, \mathbf{X}_n$ are a set of linearly independent vectors. Let

$$\mathbf{Y} = c_1 \mathbf{X}_1 + c_2 \mathbf{X}_2 \ldots + c_n \mathbf{X}_n \tag{4.7}$$

It is now required to show that scalars $c_1, c_2, \ldots, c_n$ cannot be found such that $\mathbf{Y} = \mathbf{0}$.

Multiply both sides of equation (4.7) by $(\mathbf{A} - \lambda_i \mathbf{I})$, where λ_i is any eigenvalue, then

$$(\mathbf{A} - \lambda_i \mathbf{I})\mathbf{Y} = c_1 \mathbf{A} \mathbf{X}_1 + c_2 \mathbf{A} \mathbf{X}_2 \ldots + c_n \mathbf{A}\mathbf{X}_n - c_1\lambda_i\mathbf{X}_1 - c_2\lambda_i\mathbf{X}_2 \ldots - c_n\lambda_i \mathbf{X}_n$$

But, $\mathbf{X}_1 \ldots \mathbf{X}_n$ are all eigenvectors and so

$$\mathbf{A} \mathbf{X}_j = \lambda_j \mathbf{X}_j, \text{ for all } j,$$

giving $(\mathbf{A} - \lambda_i \mathbf{I})\mathbf{Y} = c_1(\lambda_1 - \lambda_i)\mathbf{X}_1 + c_2(\lambda_2 - \lambda_i)\mathbf{X}_2 \ldots$

$$c_n(\lambda_n - \lambda_i)\mathbf{X}_n \qquad (4.8)$$

The term involving $c_i \mathbf{X}_i$ has therefore been eliminated from equation (4.8). Repeating this type of multiplication, we obtain

$$(\mathbf{A} - \lambda_1 \mathbf{I})(\mathbf{A} - \lambda_2 \mathbf{I}) \ldots (\mathbf{A} - \lambda_{n-1}\mathbf{I})\mathbf{Y} = (\lambda_1 - \lambda_n)(\lambda_2 - \lambda_n) \ldots (\lambda_{n-1} - \lambda_n) c_n \mathbf{X}_n \qquad (4.9)$$

and, since all the eigenvalues are distinct, $\mathbf{Y} = \mathbf{0}$ only if $c_n = 0$. However, a similar multiplication to that in equation (4.9) can be carried out for every eigenvalue and so $\mathbf{Y} = \mathbf{0}$ only if $c_1 = c_2 = \ldots = c_n = 0$, which shows that $\mathbf{X}_1, \mathbf{X}_2, \ldots, \mathbf{X}_n$ form a linearly independent set of vectors.

Example 4.2: Reduce the matrix $\mathbf{A}$, with $\mathbf{A}$ given below, to a diagonal matrix.

$$\mathbf{A} = \begin{pmatrix} 0 & 0 & 1 \\ 3 & 7 & -9 \\ 0 & 2 & -1 \end{pmatrix}$$

$$|\mathbf{A} - \lambda \mathbf{I}| = \begin{vmatrix} -\lambda & 0 & 1 \\ 3 & 7-\lambda & -9 \\ 0 & 2 & -1-\lambda \end{vmatrix} = 0$$

On evaluating this determinant, the characteristic equation is given by

$$\lambda^3 - 6\lambda^2 + 11\lambda - 6 = 0$$

which can easily be factorized to give

$$(\lambda - 1)(\lambda - 2)(\lambda - 3) = 0$$

The eigenvalues therefore are 1, 2 and 3, with a sum of 6.
When $\lambda = 1$,

$$(\mathbf{A}-\lambda\mathbf{I})\,\mathbf{X} = \begin{pmatrix} -1 & 0 & 1 \\ 3 & 6 & -9 \\ 0 & 2 & -2 \end{pmatrix}\begin{pmatrix} x_1 \\ x_2 \\ x_3 \end{pmatrix} = 0$$

The solution to this homogeneous set of equations is clearly

$$\begin{pmatrix} x_1 \\ x_2 \\ x_3 \end{pmatrix} = \mu\begin{pmatrix} 1 \\ 1 \\ 1 \end{pmatrix}$$

giving $\begin{pmatrix} 1 \\ 1 \\ 1 \end{pmatrix}$ as one eigenvector on selecting $\mu = 1$.

When $\lambda = 2$, $(\mathbf{A}-\lambda\mathbf{I})\,\mathbf{X} = \begin{pmatrix} -2 & 0 & 1 \\ 3 & 5 & -9 \\ 0 & 2 & -3 \end{pmatrix}\begin{pmatrix} x_1 \\ x_2 \\ x_3 \end{pmatrix} = \mathbf{0}$, which has a solution

$$\begin{pmatrix} x_1 \\ x_2 \\ x_3 \end{pmatrix} = \begin{pmatrix} 1 \\ 3 \\ 2 \end{pmatrix}$$

on selecting $\mu = 1$, giving the second eigenvector.

When $\lambda = 3$, $(\mathbf{A}-\lambda\mathbf{I})\,\mathbf{X} = \begin{pmatrix} -3 & 0 & 1 \\ 3 & 4 & -9 \\ 0 & 2 & -4 \end{pmatrix}\begin{pmatrix} x_1 \\ x_2 \\ x_3 \end{pmatrix} = \mathbf{0}$

giving a third eigenvector as $\begin{pmatrix} 1 \\ 6 \\ 3 \end{pmatrix}$

The matrix $\mathbf{Q}$ is therefore given by

$$\mathbf{Q} = \begin{pmatrix} 1 & 1 & 1 \\ 1 & 3 & 6 \\ 1 & 2 & 3 \end{pmatrix}$$

The inverse of this matrix is given by

$$\mathbf{Q}^{-1} = \begin{pmatrix} 3 & 1 & -3 \\ -3 & -2 & 5 \\ 1 & 1 & -2 \end{pmatrix}$$

and

$$\mathbf{Q}^{-1}\mathbf{A}\,\mathbf{Q} = \begin{pmatrix} 3 & 1 & -3 \\ -3 & -2 & 5 \\ 1 & 1 & -2 \end{pmatrix}\begin{pmatrix} 0 & 0 & 1 \\ 3 & 7 & -9 \\ 0 & 2 & -1 \end{pmatrix}\begin{pmatrix} 1 & 1 & 1 \\ 1 & 3 & 6 \\ 1 & 2 & 3 \end{pmatrix}$$

$$= \begin{pmatrix} 3 & 1 & -3 \\ -3 & -2 & 5 \\ 1 & 1 & -2 \end{pmatrix}\begin{pmatrix} 1 & 2 & 3 \\ 1 & 6 & 18 \\ 1 & 4 & 9 \end{pmatrix} = \begin{pmatrix} 1 & 0 & 0 \\ 0 & 2 & 0 \\ 0 & 0 & 3 \end{pmatrix}$$

verifying that a diagonal matrix has been generated, a diagonal matrix that has the eigenvalues as its elements.

It is possible that independent eigenvectors may exist even though the eigenvalues are not distinct though no theorems exist to show under what conditions they do so. The two following examples illustrate the two situations that can arise.

Example 4.3: If possible reduce the matrix $\mathbf{A}$, where

$$\mathbf{A} = \begin{pmatrix} 0 & 4 \\ 0 & 0 \end{pmatrix}$$

to a diagonal matrix.

$$|\mathbf{A} - \lambda\mathbf{I}| = \begin{vmatrix} 0-\lambda & 4 \\ 0 & 0-\lambda \end{vmatrix} = 0, \text{ or } \lambda = 0$$

which shows that there is one repeated eigenvalue $\lambda = 0$.

When $\lambda = 0$, $(\mathbf{A} - \lambda\mathbf{I})\,\mathbf{X} = \begin{pmatrix} 0 & 4 \\ 0 & 0 \end{pmatrix}\begin{pmatrix} x_1 \\ x_2 \end{pmatrix} = \mathbf{0}$, and only one

independent eigenvector $\mu\begin{pmatrix} 0 \\ 1 \end{pmatrix}$ can be found. It is not therefore

possible to reduce this matrix to a diagonal form. This must clearly be the case, for if it were possible, the diagonal matrix

would be $\begin{pmatrix} 0 & 0 \\ 0 & 0 \end{pmatrix}$, the null matrix, and a finite matrix would have been reduced to a null matrix.

Example 4.4: If possible, reduce the matrix $\mathbf{A}$, where

$$\mathbf{A} = \begin{pmatrix} 2 & 1 & 1 \\ 2 & 3 & 2 \\ 3 & 3 & 4 \end{pmatrix}$$

to a diagonal matrix.

$$|\mathbf{A} - \lambda \mathbf{I}| = \begin{vmatrix} 2-\lambda & 1 & 1 \\ 2 & 3-\lambda & 2 \\ 3 & 3 & 4-\lambda \end{vmatrix} = 0$$

which gives the characteristic equation as

$$\lambda^3 - 9\lambda^2 + 15\lambda - 7 = 0$$

This cubic factorizes to give

$$(\lambda - 7)(\lambda - 1)^2 = 0$$

and so the eigenvalues are 7 and 1 repeated. Corresponding to

$\lambda = 7$ we have $\begin{pmatrix} -5 & 1 & 1 \\ 2 & -4 & 2 \\ 3 & 3 & -3 \end{pmatrix} \begin{pmatrix} x_1 \\ x_2 \\ x_3 \end{pmatrix} = \mathbf{0}$

giving as a possible eigenvector $\begin{pmatrix} 1 \\ 2 \\ 3 \end{pmatrix}$.

Corresponding to $\lambda = 1$ we require

$$\begin{pmatrix} 1 & 1 & 1 \\ 2 & 2 & 2 \\ 3 & 3 & 3 \end{pmatrix} \begin{pmatrix} x_1 \\ x_2 \\ x_3 \end{pmatrix} = \mathbf{0}$$

Now two independent sets of solution exist, being

$$\begin{pmatrix} x_1 \\ x_2 \\ x_3 \end{pmatrix} = \mu \begin{pmatrix} 1 \\ 0 \\ -1 \end{pmatrix} \text{ and } \nu \begin{pmatrix} 0 \\ 1 \\ -1 \end{pmatrix}$$

The other obvious solution, $\alpha\begin{pmatrix}1\\-1\\0\end{pmatrix}$, is not independent of the first two.

For this matrix a non-singular matrix $\mathbf{Q}$ can therefore be found, where

$$\mathbf{Q} = \begin{pmatrix}1 & 1 & 0\\2 & 0 & 1\\3 & -1 & -1\end{pmatrix}$$

the independence of the eigenvectors being verified as $|\mathbf{Q}| = 6$. Performing the necessary computation we have

$$\mathbf{Q}^{-1} = \frac{1}{6}\begin{pmatrix}1 & -1 & -1\\-5 & 1 & 1\\2 & -4 & 2\end{pmatrix}$$

$$\text{and } \mathbf{Q}^{-1}\mathbf{A}\mathbf{Q} = \frac{1}{6}\begin{pmatrix}1 & -1 & -1\\-5 & 1 & 1\\2 & -4 & 2\end{pmatrix}\begin{pmatrix}2 & 1 & 1\\2 & 3 & 2\\3 & 3 & 4\end{pmatrix}\begin{pmatrix}1 & 1 & 0\\2 & 0 & 1\\3 & -1 & -1\end{pmatrix}$$

$$= \frac{1}{6}\begin{pmatrix}1 & 1 & 1\\5 & -1 & -1\\-2 & 4 & -2\end{pmatrix}\begin{pmatrix}7 & 1 & 0\\14 & 0 & 1\\21 & 1 & -1\end{pmatrix} = \frac{1}{6}\begin{pmatrix}42 & 0 & 0\\0 & 6 & 0\\0 & 0 & 6\end{pmatrix}$$

$$= \begin{pmatrix}7 & 0 & 0\\0 & 1 & 0\\0 & 0 & 1\end{pmatrix}$$

For this matrix it is therefore possible to obtain a reduction to a diagonal matrix.

4.5. Orthogonal reduction to a diagonal matrix

If the matrix $\mathbf{Q}$ in the above work were to be an orthogonal matrix, it would be very much easier to calculate its inverse as this is simply its transpose. There is therefore considerable advantage in searching for situations where a reduction to the diagonal matrix using an orthogonal matrix is possible. First, note that if the matrix to be reduced is real and symmetric, then all its eigenvalues are real, for if

$$\mathbf{A}\,\mathbf{X} = \lambda\,\mathbf{X} \tag{4.10}$$

then on transposing

$$\mathbf{X}^T\mathbf{A}^T = \mathbf{X}^T\lambda = \lambda\mathbf{X}^T$$

while taking the complex conjugate gives

$$\bar{\mathbf{X}}^T\mathbf{A} = \bar{\lambda}\,\bar{\mathbf{X}}^T \tag{4.11}$$

The product $\bar{\mathbf{X}}^T\mathbf{A}\,\mathbf{X}$ can be generated in two ways, both as $(\bar{\mathbf{X}}^T\mathbf{A})\mathbf{X}$ and $\bar{\mathbf{X}}^T(\mathbf{A}\,\mathbf{X})$. On substituting from equations (4.10) and (4.11) we therefore have

$$\bar{\lambda}\,\bar{\mathbf{X}}^T\mathbf{X} = \lambda\,\bar{\mathbf{X}}^T\mathbf{X} \tag{4.12}$$

and so $\bar{\lambda} = \lambda$ showing that the eigenvalues are real. Since both $\mathbf{A}$ and λ are real, $\mathbf{X}$ must also be real and so the eigenvectors are real. If this is the case, $\mathbf{X}^T\mathbf{X}$ will always give a positive quantity and if

$$\mathbf{X}^T = \mu\,(x_1, x_2, \ldots, x_n) \text{ say,}$$

then $\mathbf{X}^T\mathbf{X} = \mu^2(x_1^2 + x_2^2 \ldots + x_n^2)$ and so by suitably selecting μ so that

$$\mu^2 = (x_1^2 + x_2^2 \ldots + x_n^2)^{-1} \tag{4.13}$$

it is always possible to have $\mathbf{X}^T\mathbf{X} = 1$. When this is the case, the eigenvectors have been NORMALIZED. Hence, one of the conditions necessary for $\mathbf{Q}$ to be an orthogonal matrix, namely that the norm of any column is unity, can always be satisfied by the eigenvectors of a symmetric matrix.

If λ_i and λ_j are two eigenvalues of the symmetric matrix, then

$$\mathbf{A}\,\mathbf{X}_i = \lambda_i\,\mathbf{X}_i, \text{ or } \mathbf{X}_i^T\mathbf{A} = \lambda_i\,\mathbf{X}_i^T \tag{4.14}$$

and

$$\mathbf{A}\,\mathbf{X}_j = \lambda_j\,\mathbf{X}_j \tag{4.15}$$

The product $\mathbf{X}_i^T\,\mathbf{A}\,\mathbf{X}_j$ can also be generated in two ways, as $(\mathbf{X}_i^T\mathbf{A})\,\mathbf{X}_j$ and as $\mathbf{X}_i^T(\mathbf{A}\,\mathbf{X}_j)$. Substitution from equations (4.14) and (4.15) gives

$$\lambda_i\,\mathbf{X}_i^T\mathbf{X}_j = \lambda_j\,\mathbf{X}_i^T\mathbf{X}_j$$

or

$$(\lambda_i - \lambda_j)\mathbf{X}_i^T\mathbf{X}_j = 0 \tag{4.16}$$

If the symmetric matrix has distinct eigenvalues so that $\lambda_i \neq \lambda_j$ then equation (4.16) implies that

$$\mathbf{X}_i{}^T \mathbf{X}_j = 0$$

The second condition which $\mathbf{Q}$ has to satisfy in order to be an orthogonal matrix is therefore also fulfilled. Hence, a real symmetric matrix with distinct eigenvalues can always be reduced to a diagonal matrix by means of an orthogonal matrix.

Example 4.5: By means of an orthogonal matrix, transform the matrix $\mathbf{A}$, where

$$\mathbf{A} = \begin{pmatrix} 11 & 2 & 8 \\ 2 & 2 & -10 \\ 8 & -10 & 5 \end{pmatrix}$$

to a diagonal matrix.

$$|\mathbf{A} - \lambda \mathbf{I}| = \begin{vmatrix} 11-\lambda & 2 & 8 \\ 2 & 2-\lambda & -10 \\ 8 & -10 & 5-\lambda \end{vmatrix} = 0$$

which generates the characteristic equation

$$\lambda^3 - 18\lambda^2 - 81\lambda + 1458 = 0$$

This equation can be factorized to give

$$(\lambda + 9)(\lambda - 9)(\lambda - 18) = 0$$

The eigenvalues are therefore -9, 9 and 18. As these are distinct and the matrix $\mathbf{A}$ is symmetric, the general theory outlined above shows that it is possible to reduce by means of an orthogonal matrix.

When $\lambda = -9$, $(\mathbf{A} - \lambda \mathbf{I})\,\mathbf{X} = \begin{pmatrix} 20 & 2 & 8 \\ 2 & 11 & 10 \\ 8 & -10 & 14 \end{pmatrix} \begin{pmatrix} x_1 \\ x_2 \\ x_3 \end{pmatrix} = \mathbf{O}$

Solving the necessary set of homogeneous equations, an eigenvector is given by

$$\mathbf{X} = \mu \begin{pmatrix} -1 \\ 2 \\ 2 \end{pmatrix}$$

Normalizing gives $\mathbf{X}^T\mathbf{X} = 9\mu^2$ and so the eigenvector we require is

$$\frac{1}{3}\begin{pmatrix} -1 \\ 2 \\ 2 \end{pmatrix}$$

When $\lambda = 9$,

$$(\mathbf{A} - \lambda\,\mathbf{I})\,\mathbf{X} = \begin{pmatrix} 2 & 2 & 8 \\ 2 & -7 & -10 \\ 8 & -10 & -4 \end{pmatrix}\begin{pmatrix} x_1 \\ x_2 \\ x_3 \end{pmatrix} = \mathbf{O}$$

and an eigenvector now is

$$\mathbf{X} = \mu\begin{pmatrix} 2 \\ 2 \\ -1 \end{pmatrix}$$

Normalizing this vector gives the eigenvector we require as

$$\frac{1}{3}\begin{pmatrix} 2 \\ 2 \\ -1 \end{pmatrix}$$

When $\lambda = 18$,

$$(\mathbf{A} - \lambda\,\mathbf{I})\,\mathbf{X} = \begin{pmatrix} -9 & 2 & 8 \\ 2 & -16 & -10 \\ 8 & -10 & -13 \end{pmatrix}\begin{pmatrix} x_1 \\ x_2 \\ x_3 \end{pmatrix} = \mathbf{O}$$

and an eigenvector is now given by

$$\mu\begin{pmatrix} 2 \\ -1 \\ 2 \end{pmatrix}$$

Normalizing this vector, the required eigenvector is

$$\frac{1}{3}\begin{pmatrix} 2 \\ -1 \\ 2 \end{pmatrix}$$

The matrix of transformation is therefore given by

$$\mathbf{Q} = \frac{1}{3}\begin{pmatrix} -1 & 2 & 2 \\ 2 & 2 & -1 \\ 2 & -1 & 2 \end{pmatrix}$$

and it is easy to verify that this is an orthogonal matrix, while

$$\mathbf{Q}^{-1}\mathbf{A}\mathbf{Q} = \mathbf{Q}^{T}\mathbf{A}\mathbf{Q} =$$

$$\frac{1}{9}\begin{pmatrix} -1 & 2 & 2 \\ 2 & 2 & -1 \\ 2 & -1 & 2 \end{pmatrix}\begin{pmatrix} 11 & 2 & 8 \\ 2 & 2 & -1 \\ 8 & -10 & 5 \end{pmatrix}\begin{pmatrix} -1 & 2 & 2 \\ 2 & 2 & -1 \\ 2 & -1 & 2 \end{pmatrix}$$

$$= \begin{pmatrix} -9 & 0 & 0 \\ 0 & 9 & 0 \\ 0 & 0 & 18 \end{pmatrix}$$

It is in fact possible to prove that even if a symmetric matrix has repeated eigenvalues, the correct number of eigenvectors can be found and that these eigenvectors can be selected so that they form an orthogonal matrix. The proof that the correct number of eigenvectors exist is long and has been omitted while the procedure for generating an orthogonal set from the eigenvectors found is best illustrated by means of an example.

Example 4.6: Reduce the matrix $\mathbf{A}$, where

$$\mathbf{A} = \begin{pmatrix} 5 & 2 & 2 \\ 2 & 2 & 1 \\ 2 & 1 & 2 \end{pmatrix}$$

to a diagonal matrix using an orthogonal matrix.

$$|\mathbf{A} - \lambda\,\mathbf{I}| = \begin{vmatrix} 5-\lambda & 2 & 2 \\ 2 & 2-\lambda & 1 \\ 2 & 1 & 2-\lambda \end{vmatrix} = 0,$$ which gives the characteristic equation as

$$\lambda^3 - 9\lambda^2 + 15\lambda - 7 = 0$$

On factorizing, the eigenvalues are found to be 7 and 1 repeated. When $\lambda = 7$,

$$(\mathbf{A} - \lambda\,\mathbf{I})\,\mathbf{X} = \begin{pmatrix} -2 & 2 & 2 \\ 2 & -5 & 1 \\ 2 & 1 & -5 \end{pmatrix}\begin{pmatrix} x \\ x \\ x \end{pmatrix} = \mathbf{0}$$

and a possible eigenvector, on solving, is given by

$$\mathbf{X} = \mu \begin{pmatrix} 2 \\ 1 \\ 1 \end{pmatrix}$$

$\mathbf{X}^T\mathbf{X} = 6\mu^2$ and so, on normalizing, the required eigenvector is

$$\mathbf{X} = \frac{1}{\sqrt{6}} \begin{pmatrix} 2 \\ 1 \\ 1 \end{pmatrix}$$

When $\lambda = 1$,

$$(\mathbf{A} - \lambda\,\mathbf{I})\,\mathbf{X} = \begin{pmatrix} 4 & 2 & 2 \\ 2 & 1 & 1 \\ 2 & 1 & 1 \end{pmatrix} \begin{pmatrix} x_1 \\ x_2 \\ x_3 \end{pmatrix} = \mathbf{0}$$

Now two independent families of solutions can be found to these homogeneous equations, namely,

$$\mu \begin{pmatrix} -1 \\ 2 \\ 0 \end{pmatrix} \quad \text{and} \quad \nu \begin{pmatrix} 0 \\ 1 \\ -1 \end{pmatrix}$$

while any linear combination of these two, namely

$$\alpha \begin{pmatrix} -1 \\ 2 \\ 0 \end{pmatrix} + \beta \begin{pmatrix} 0 \\ 1 \\ -1 \end{pmatrix}$$

will also be an eigenvector.

Both families are orthogonal to the first eigenvector found but they are not orthogonal to one another. Arbitrarily selcting $\mu \begin{pmatrix} -1 \\ 2 \\ 0 \end{pmatrix}$ to be the second eigenvector when it has been normalized, that is $\frac{1}{\sqrt{5}} \begin{pmatrix} -1 \\ 2 \\ 0 \end{pmatrix}$, the third eigenvector must be selected to be orthogonal to this, that is the values of α and β must be selected so that

$$\frac{1}{\sqrt{5}}(-1\ 2\ 0)\left(\alpha\begin{pmatrix}-1\\2\\0\end{pmatrix}+\beta\begin{pmatrix}0\\1\\-1\end{pmatrix}\right)=0$$

or $5\alpha + 2\beta = 0$, that is $\beta = -5\alpha/2$. A possible third eigenvector is therefore

$$\alpha\left(\begin{pmatrix}-1\\2\\0\end{pmatrix}-\frac{5}{2}\begin{pmatrix}0\\1\\-1\end{pmatrix}\right)=\alpha\begin{pmatrix}-1\\-1/2\\5/2\end{pmatrix}$$

Normalizing requires $\mathbf{X}^T\mathbf{X} = \dfrac{30\alpha}{4}$ and so the third eigenvector is $\dfrac{1}{\sqrt{30}}\begin{pmatrix}2\\1\\-5\end{pmatrix}$, while the transforming matrix, $\mathbf{Q}$, is

$$\mathbf{Q}=\frac{1}{\sqrt{30}}\begin{pmatrix}2\sqrt{5} & -\sqrt{6} & 2\\ \sqrt{5} & 2\sqrt{6} & 1\\ \sqrt{5} & 0 & -5\end{pmatrix}$$

which is an orthogonal matrix.

$$\mathbf{Q}^{-1}\mathbf{A}\,\mathbf{Q}=\mathbf{Q}^T\mathbf{A}\,\mathbf{Q}=\frac{1}{30}\begin{pmatrix}2\sqrt{5} & \sqrt{5} & \sqrt{5}\\ -\sqrt{6} & 2\sqrt{6} & 0\\ 2 & 1 & -5\end{pmatrix}\begin{pmatrix}5 & 2 & 2\\ 2 & 2 & 1\\ 2 & 1 & 2\end{pmatrix}\begin{pmatrix}2\sqrt{5} & -\sqrt{6} & 2\\ \sqrt{5} & 2\sqrt{6} & 1\\ \sqrt{5} & 0 & -5\end{pmatrix}$$

$$=\begin{pmatrix}7 & 0 & 0\\ 0 & 1 & 0\\ 0 & 0 & 1\end{pmatrix}$$

4.6. Geometric application

Although, as a matter of general policy, the application of matrices to specific areas has been left until Chapter 6, the insertion of this particular application here will facilitate the understanding of the Jacobi method for determining the eigenvalues which is to be described in Chapter 5.

Let $\mathbf{Q}$ be an orthogonal matrix of order 3, where

$$\mathbf{Q}=\begin{pmatrix}l_1 & l_2 & l_3\\ m_1 & m_2 & m_3\\ n_1 & n_2 & n_3\end{pmatrix}$$

Then

$$\begin{aligned} l_1^2 + m_1^2 + n_1^2 &= 1 \\ l_2^2 + m_2^2 + n_2^2 &= 1 \\ l_3^2 + m_3^2 + n_3^2 &= 1 \end{aligned} \qquad (4.17)$$

and

$$\begin{aligned} l_1 l_2 + m_1 m_2 + n_1 n_2 &= 0 \\ l_1 l_3 + m_1 m_3 + n_1 n_3 &= 0 \\ l_3 l_2 + m_3 m_2 + n_3 n_2 &= 0 \end{aligned} \qquad (4.18)$$

However, equations (4.17) and (4.18) are exactly the equations which the direction cosines of one set of axes, $Ox'y'z'$ say, would satisfy when referred to another set, $Oxyz$. An orthogonal matrix of order 3 can therefore always be regarded as a matrix which rotates one set of cartesian coordinate axes to another set. An orthogonal matrix or order 2 is of course a special case of this and can in fact be written as

$$\begin{pmatrix} \cos\theta & \sin\theta \\ -\sin\theta & \cos\theta \end{pmatrix}$$

where θ is the angle through which one axis has been rotated.

A general quadric (i.e., ellipsoid paraboloid, etc.) with a suitable choice of origin can always be written in the form

$$ax^2 + by^2 + cz^2 + 2fyz + 2gzx + 2hxy = d$$

or

$$(x \;\; y \;\; z)\begin{pmatrix} a & h & g \\ h & b & f \\ g & f & c \end{pmatrix}\begin{pmatrix} x \\ y \\ z \end{pmatrix} = d \qquad (4.19)$$

where the matrix, $\mathbf{A}$ say, is symmetric. Such a matrix can therefore be reduced to a diagonal matrix by a transformation of the form $\mathbf{Q}^T\mathbf{A}\,\mathbf{Q}$, where $\mathbf{Q}$ is an orthogonal matrix. Hence by suitably rotating the axes so that

$$\begin{pmatrix} x \\ y \\ z \end{pmatrix} = \mathbf{Q}\begin{pmatrix} x' \\ y' \\ z' \end{pmatrix}$$

equation (4.19) becomes

$$(x' \; y' \; z')\, \mathbf{Q}^T \mathbf{A}\, \mathbf{Q} \begin{pmatrix} x' \\ y' \\ z' \end{pmatrix} = \mathbf{X}'\, \Lambda\, \mathbf{X}' = d \tag{4.20}$$

where Λ is a diagonal matrix. Equation (4.20) gives

$$\lambda_1 x'^2 + \lambda_2 y'^2 + \lambda_3 z'^2 = d$$

the equation of a quadric with its principal axes as a coordinate system.

For matrices of order 2 and 3, the orthogonal reduction to a diagonal matrix therefore corresponds to a suitable rotation of the coordinate system until the coordinate axes are the principal axes of the quadric.

Example 4.7: Show that the quadric

$$5x^2 + 6y^2 + 7z^2 - 4xy + 4yz = 162$$

is an ellipsoid.

The quadric may be written as

$$(x \;\; y \;\; z) \begin{pmatrix} 5 & -2 & 0 \\ -2 & 6 & 2 \\ 0 & 2 & 7 \end{pmatrix} \begin{pmatrix} x \\ y \\ z \end{pmatrix} = 162$$

Considering the matrix only we have

$$|\mathbf{A} - \lambda \mathbf{I}| = \begin{vmatrix} 5-\lambda & -2 & 0 \\ -2 & 6-\lambda & 2 \\ 0 & 2 & 7-\lambda \end{vmatrix} = 0,$$ which gives a characteristic equation as

$$\lambda^3 - 18\lambda^2 + 99\lambda - 162 = 0$$

which factorizes to $(\lambda - 3)(\lambda - 6)(\lambda - 9) = 0$, giving the eigenvalues as 3, 6 and 9.

When $\lambda = 3$,

$$(\mathbf{A} - \lambda \mathbf{I})\, \mathbf{X} = \begin{pmatrix} 2 & -2 & 0 \\ -2 & 3 & 2 \\ 0 & 2 & 4 \end{pmatrix} \begin{pmatrix} x_1 \\ x_2 \\ x_3 \end{pmatrix} = \mathbf{0}$$

giving as a normalized eigenvector

$$\frac{1}{3}\begin{pmatrix} 2 \\ 2 \\ -1 \end{pmatrix}$$

When $\lambda = 6$,

$$(\mathbf{A} - \lambda \mathbf{I})\,\mathbf{X} = \begin{pmatrix} -1 & -2 & 0 \\ -2 & 0 & 2 \\ 0 & 2 & 1 \end{pmatrix}\begin{pmatrix} x_1 \\ x_2 \\ x_3 \end{pmatrix} = \mathbf{O}$$

giving as a normalized eigenvector

$$\frac{1}{3}\begin{pmatrix} 2 \\ -1 \\ 2 \end{pmatrix}$$

When $\lambda = 9$,

$$(\mathbf{A} - \lambda \mathbf{I})\,\mathbf{X} = \begin{pmatrix} -4 & -2 & 2 \\ -2 & -3 & 2 \\ 0 & 2 & -2 \end{pmatrix}\begin{pmatrix} x_1 \\ x_2 \\ x_3 \end{pmatrix} = \mathbf{O}$$

giving as a normalized eigenfactor

$$\frac{1}{3}\begin{pmatrix} -1 \\ 2 \\ 2 \end{pmatrix}$$

Hence, if $\mathbf{Q} = \frac{1}{3}\begin{pmatrix} 2 & 2 & -1 \\ 2 & -1 & 2 \\ -1 & 2 & 2 \end{pmatrix}$, then

$$\mathbf{Q}^T\mathbf{A}\,\mathbf{Q} = \begin{pmatrix} 3 & 0 & 0 \\ 0 & 6 & 0 \\ 0 & 0 & 9 \end{pmatrix}$$

Select therefore a new reference set of axis given by

$\begin{pmatrix} x \\ y \\ z \end{pmatrix} = \mathbf{Q}\begin{pmatrix} x' \\ y' \\ z' \end{pmatrix}$, and the quadric equation becomes

$$(x'\,y'\,z')\,\mathbf{Q}^T\mathbf{A}\,\mathbf{Q}\begin{pmatrix}x'\\y'\\z'\end{pmatrix} = (x'\,y'\,z')\begin{pmatrix}3&0&0\\0&6&0\\0&0&9\end{pmatrix}\begin{pmatrix}x'\\y'\\z'\end{pmatrix} = 162$$

or
$$3x'^2 + 6y'^2 + 9z'^2 = 162$$

which, when written in the form

$$\frac{x'^2}{54} + \frac{y'^2}{27} + \frac{z'^2}{18} = 1$$

is recognizable as an ellipsoid.

4.7. Spectral resolution

As previously mentioned, it is possible to find a row eigenvector corresponding to every column eigenvector. Let $\mathbf{R}$ be the matrix with the row eigenvectors as its rows, which corresponds to $\mathbf{Q}$, the matrix with the column eigenvectors as its columns. Also let $\mathbf{Y}_i^T$ denote the row vector corresponding to the column vector $\mathbf{X}_i$.

Now
$$\mathbf{Y}_i^T\,\mathbf{A} = \lambda_i\mathbf{Y}_i^T$$

and
$$\mathbf{A}\,\mathbf{X}_j = \lambda_j\mathbf{X}_j$$

The product $\mathbf{Y}_i^T\,\mathbf{A}\,\mathbf{X}_j$ can therefore be generated in two ways, giving

$$\mathbf{Y}_i^T(\mathbf{A}\,\mathbf{X}_j) = \lambda_j\mathbf{Y}_i^T\,\mathbf{X}_j \quad \text{and} \quad (\mathbf{Y}_i^T\,\mathbf{A})\,\mathbf{X}_j = \lambda_i\mathbf{Y}_i^T\,\mathbf{X}_j$$

For consistency we must have

$$(\lambda_i - \lambda_j)\,\mathbf{Y}_i^T\mathbf{X}_j = 0 \tag{4.21}$$

so that if all the eigenvalues are distinct, the row eigenvector corresponding to one eigenvalue is orthogonal to the column eigenvector corresponding to another eigenvalue. Further, if a normalization process which makes

$$\mathbf{Y}_i^T\mathbf{X}_i = 1 \tag{4.22}$$

is adopted, then

$$\mathbf{R}\,\mathbf{Q} = \mathbf{I} \text{ or } \mathbf{R} = \mathbf{Q}^{-1} \text{ and } \mathbf{Q} = \mathbf{R}^{-1}$$

If Λ is the matrix of eigenvalues, then

$$\mathbf{A}\,\mathbf{Q} = \mathbf{Q}\,\Lambda$$

and
$$\mathbf{R}\,\mathbf{A} = \Lambda\,\mathbf{R}$$

in view of the construction of **R** and **Q**. Hence

$$\mathbf{R\,A\,Q} = \mathbf{R\,Q}\Lambda = \mathbf{I}\,\Lambda = \Lambda$$

or, in a more useful way

$$\mathbf{A} = \mathbf{R}^{-1}\Lambda\,\mathbf{Q}^{-1} = \mathbf{Q}\,\Lambda\,\mathbf{R} \tag{4.23}$$

Remembering that Λ is a diagonal matrix, equation (4.23) gives

$$\mathbf{A} = \mathbf{X}_1\mathbf{Y}_1^T\lambda_1 + \mathbf{X}_2\mathbf{Y}_2^T\lambda_2 \ldots \mathbf{X}_n\mathbf{Y}_n^T\lambda_n \tag{4.23}$$

This expansion is called the spectral resolution of the matrix **A**, the terminology arising out of the physical case when the eigenvalues are regarded as frequencies and so equation (4.23) resolves the matrix into its constituent frequency. Note that since $\mathbf{Y}_i^T\,\mathbf{X}_j = 0$ and $\mathbf{Y}_i^T\,\mathbf{X}_i = 1$

$$\mathbf{A}^2 = \mathbf{X}_1\mathbf{Y}_1^T\lambda_1^2 + \mathbf{X}_2\,\mathbf{Y}_2^T\lambda_2^2 \ldots + \mathbf{X}_n\mathbf{T}_n^T\lambda_n^2$$

and

$$\mathbf{A}^m = \mathbf{X}_1\,\mathbf{Y}_1^T\lambda_1^m + \mathbf{X}_2\,\mathbf{Y}_2^T\lambda_2^m \ldots + \mathbf{X}_n\mathbf{Y}_n^T\lambda_n^m$$

so that if λ_1 say is the dominant eigenvalue, for sufficiently large m

$$\mathbf{A}^m \simeq \mathbf{X}_1\mathbf{Y}_1^T\lambda_1^m \tag{4.24}$$

If **A** is a symmetric matrix, then $\mathbf{X} = \mathbf{Y}$ and so

$$\mathbf{R}^T = \mathbf{Q} \text{ and } \mathbf{R} = \mathbf{Q}^T$$

However, $\mathbf{Q} = \mathbf{R}^{-1}$ and $\mathbf{R} = \mathbf{Q}^{-1}$, which shows that $\mathbf{R}^T = \mathbf{R}^{-1}$ and $\mathbf{Q}^T = \mathbf{Q}^{-1}$, thus verifying again that the matrix of eigenvectors of a symmetric matrix can be an orthogonal matrix. For this case, equations (4.23) and (4.24) become

$$\mathbf{A} = \mathbf{X}_1\mathbf{X}_1^T\lambda_1 + \mathbf{X}_2\mathbf{X}_2^T\lambda_2 \ldots \mathbf{X}_n\mathbf{X}_n^T\lambda_n \tag{4.25}$$

and

$$\mathbf{A}^m \simeq \mathbf{X}_1\mathbf{X}_1^T\lambda_1^m \tag{4.26}$$

4.8. Functions of a matrix

Let **A** be any square matrix with an eigenvalue λ and a corresponding eigenvector **X** so that

$$\mathbf{A\,X} = \lambda\,\mathbf{X}$$

Then

$$\mathbf{A}^2\mathbf{X} = \mathbf{A}\,\lambda\,\mathbf{X} = \lambda\,\mathbf{A\,X} = \lambda^2\,\mathbf{X}$$

Similarly $\mathbf{A}^m\mathbf{X} = \lambda^m\mathbf{X}$ where m is any integer. Hence, if $f(\mathbf{A})$ is any function of the matrix $\mathbf{A}$ which involves only integer powers, then

$$f(\mathbf{A})\,\mathbf{X} = f(\lambda)\,\mathbf{X} \tag{4.27}$$

so that the function of a matrix is replaced by a scalar function.

A useful special case arises in connection with the set of equations

$$(\mathbf{A} - \mu\mathbf{I})\,\mathbf{Y} = \mathbf{X}$$

from which $\mathbf{Y}$ is to be determined, so that $\mathbf{Y} = (\mathbf{A} - \mu\mathbf{I})^{-1}\mathbf{X}$. Applying equation (4.27) gives

$$\mathbf{Y} = (\lambda - \mu)^{-1}\mathbf{X}$$

which is a very simple calculation for the solution.

4.9. Exponential of a matrix

In many physical problems whose solution depends on integrating differential equations of the first order the concept of the exponential of a matrix, $e^{\mathbf{A}}$, is very useful. An example of such a problem will be given in Chapter 6. The exponential can be defined by means of a power series in the usual way, that is

$$e^{\mathbf{A}} = \mathbf{I} + \mathbf{A} + \frac{1}{2}\mathbf{A}^2 + \frac{1}{3!}\mathbf{A}^3 \ldots + \frac{1}{r!}\mathbf{A}^r \ldots \tag{4.28}$$

In many cases it will be totally impractical to evaluate this series numerically as very large powers of $\mathbf{A}$ will have to be calculated if a sufficient number of terms are to be included. On using the theorem expressed in equation (4.27), considerable manipulation can be avoided, for if $\mathbf{X}_i$ is the eigenvector corresponding to the eigenvalue λ_i

$$e^{\mathbf{A}}\mathbf{X}_i = e^{\lambda_i}\mathbf{X}_i \tag{4.29}$$

with a similar expression for every value of i. Thus, combining all these equations we have

$$e^{\mathbf{A}}\mathbf{Q} = \mathbf{Q}\,\mathbf{exp}\Lambda \tag{4.30}$$

where $\mathbf{exp}\Lambda$ is the diagonal matrix whose elements are the exponential of successive eigenvalues.

On rearranging equation (4.30) we have

$$e^{\mathbf{A}} = \mathbf{Q} \exp\Lambda \; \mathbf{Q}^{-1} \tag{4.31}$$

which is a much simpler calculation than that embodied in equation (4.28) as the exponential of scalars are already tabulated.

Example 4.8: Evaluate $e^{\mathbf{A}}$ where

$$\mathbf{A} = \begin{pmatrix} 0{\cdot}1 & 0{\cdot}1 \\ 0 & 0{\cdot}2 \end{pmatrix}$$

In order to use equation (4.31) the eigenvalues and corresponding eigenvectors of $\mathbf{A}$ must be found.

$$|\mathbf{A} - \lambda \mathbf{I}| = \begin{vmatrix} 0{\cdot}1 - \lambda & 0{\cdot}1 \\ 0 & 0{\cdot}2 - \lambda \end{vmatrix} = 0$$

The characteristic equation is therefore $(0{\cdot}1 - \lambda)\,(0{\cdot}2 - \lambda) = 0$ which gives the eigenvalues as $0{\cdot}1$ and $0{\cdot}2$.
When $\lambda = 0{\cdot}1$

$$(\mathbf{A} - \lambda \mathbf{I})\, \mathbf{X} = \begin{pmatrix} 0 & 0{\cdot}1 \\ 0 & 0{\cdot}1 \end{pmatrix} \begin{pmatrix} x_1 \\ x_2 \end{pmatrix}$$

with an obvious family $\mu \begin{pmatrix} 1 \\ 0 \end{pmatrix}$, the simplest being $\begin{pmatrix} 1 \\ 0 \end{pmatrix}$.

When $\lambda = 0.2$,

$$(\mathbf{A} - \lambda \mathbf{I})\, \mathbf{X} = \begin{pmatrix} -0{\cdot}1 & 0{\cdot}1 \\ 0 & 0 \end{pmatrix} \begin{pmatrix} x_1 \\ x_2 \end{pmatrix}$$

again an obvious eigenvector family being $\mu \begin{pmatrix} 1 \\ 1 \end{pmatrix}$, the simplest being $\begin{pmatrix} 1 \\ 1 \end{pmatrix}$. The matrix of eigenvectors is therefore

$$\mathbf{Q} = \begin{pmatrix} 1 & 0 \\ 0 & 1 \end{pmatrix}$$

Its inverse is the matrix

$$\mathbf{Q}^{-1} = \begin{pmatrix} 1 & -1 \\ 0 & 1 \end{pmatrix}$$

Thus, as $e^{\mathbf{A}} = \mathbf{Q} \exp \Lambda \; \mathbf{Q}^{-1}$,

$$e^{\mathbf{A}} = \begin{pmatrix} 1 & 1 \\ 0 & 1 \end{pmatrix} \begin{pmatrix} e^{0{\cdot}1} & 0 \\ 0 & e^{0{\cdot}2} \end{pmatrix} \begin{pmatrix} 1 & -1 \\ 0 & 1 \end{pmatrix}$$

$$= \begin{pmatrix} 1 & 1 \\ 0 & 1 \end{pmatrix} \begin{pmatrix} 1{\cdot}105 & 0 \\ 0 & 0{\cdot}221 \end{pmatrix} \begin{pmatrix} 1 & -1 \\ 0 & 1 \end{pmatrix} = \begin{pmatrix} 1 & 1 \\ 0 & 1 \end{pmatrix} \begin{pmatrix} 1{\cdot}105 & -1{\cdot}105 \\ 0 & 1{\cdot}221 \end{pmatrix}$$

$$e^{\mathbf{A}} = \begin{pmatrix} 1{\cdot}105 & 0{\cdot}116 \\ 0 & 1{\cdot}221 \end{pmatrix}$$

Evaluating from the power series, we have

$$\mathbf{A}^2 = \begin{pmatrix} 0{\cdot}1 & 0{\cdot}1 \\ 0 & 0{\cdot}2 \end{pmatrix} \begin{pmatrix} 0{\cdot}1 & 0{\cdot}1 \\ 0 & 0{\cdot}2 \end{pmatrix} = \begin{pmatrix} 0{\cdot}01 & 0{\cdot}03 \\ 0 & 0{\cdot}04 \end{pmatrix}$$

and $$\mathbf{A}^3 = \begin{pmatrix} 0{\cdot}1 & 0{\cdot}1 \\ 0 & 0{\cdot}2 \end{pmatrix} \begin{pmatrix} 0{\cdot}001 & 0{\cdot}03 \\ 0 & 0{\cdot}04 \end{pmatrix} = \begin{pmatrix} 0{\cdot}001 & 0{\cdot}007 \\ 0 & 0{\cdot}008 \end{pmatrix}$$

Then $$e^{\mathbf{A}} = \mathbf{I} + \mathbf{A} + \tfrac{1}{2}\mathbf{A}^2 + \tfrac{1}{6}\mathbf{A}^3 = \begin{pmatrix} 1{\cdot}105 & 0{\cdot}116 \\ 0 & 1{\cdot}221 \end{pmatrix}$$

so that three terms have to be included to obtain the four figure accuracy (dictated by evaluating e^{λ} to four figures) of the first method. For the first method the computational work is approximately the same for any order of accuracy (more accurate determination of e^{λ} only is required) while the evaluation by series requires the inclusion of more and more terms.

EXERCISES 4

1. Find the eigenvalues and corresponding eigenvectors of the following matrices

$$\begin{pmatrix} -1 & 0 & 2 \\ 0 & 1 & 2 \\ 2 & 2 & 0 \end{pmatrix}, \begin{pmatrix} -5 & -25 & 6 \\ 0 & 2 & 0 \\ -3 & -9 & 4 \end{pmatrix}, \begin{pmatrix} 8 & -12 & 15 \\ 15 & -25 & 33 \\ 8 & -14 & 19 \end{pmatrix}$$

Hence reduce them to diagonal matrices.

2. Use an orthogonal transformation to reduce the following matrices to diagonal matrices.

$$\begin{pmatrix} 0 & 0 & 2 \\ 0 & 0 & 0 \\ 2 & 0 & 1 \end{pmatrix}, \begin{pmatrix} -3 & 0 & 6 \\ 0 & 3 & 6 \\ 6 & 6 & 0 \end{pmatrix}, \begin{pmatrix} -1 & 2 & 2 \\ 2 & -1 & 2 \\ 2 & 2 & -1 \end{pmatrix}$$

3. A quadric has the equation

$$x^2 + 5y^2 + 3z^2 - 8xz + 8yz = -81$$

Express its equation in the coordinate reference frame coincident with its principal axes.

4. Obtain the spectral resolution of the matrix

$$\begin{pmatrix} 5 & -2 & 0 \\ -2 & 6 & 2 \\ 0 & 2 & 7 \end{pmatrix}$$

5. Evaluate $e^{\mathbf{A}}$ where $\mathbf{A} = \begin{pmatrix} 0{\cdot}1 & 0{\cdot}4 \\ 0 & 0{\cdot}2 \end{pmatrix}$

5

NUMERICAL DETERMINATION OF EIGENVALUES AND EIGENVECTORS

5.1. Introduction

In general a solution can only be obtained to a polynomial of degree less than or equal to three. This means that for matrices or order 4 or more, even when the characteristic equation has been obtained, it cannot be solved and so the eigenvalues cannot be determined by this method. In this chapter we describe some of the methods that are available. They are not of necessity the most suitable methods for all the problems that may occur, but have been selected because they are easy to follow while at the same time do work quite well. Programs will be found to accommodate them in most computer libraries, an important consideration as the developing of a new program is a very time-consuming occupation. Throughout this chapter, the largest (smallest) eigenvalue will be taken to mean largest (smallest) in absolute magnitude.

5.2. The arbitrary vector method for determining the largest eigenvalue

If a matrix $\mathbf{A}$ of order n has n linearly independent eigenvectors $\mathbf{X}_1, \mathbf{X}_2, \ldots, \mathbf{X}_n$, then any other vector $\mathbf{V}_0$, also of order n, can be expressed as a linear combination of the eigenvectors, that is

$$\mathbf{V}_0 = c_1\mathbf{X}_1 + c_2\mathbf{X}_2 \ldots + c_n\mathbf{X}_n \tag{5.1}$$

where some, if not all, of the scalar constants $c_1, c_2, \ldots, c_n$ are non-zero.

Multiplying equation (5.1) by the matrix $\mathbf{A}$ gives

$$\mathbf{V}_1 = \mathbf{A}\,\mathbf{V}_0 = c_1\mathbf{A}\,\mathbf{X}_1 + c_2\,\mathbf{A}\,\mathbf{X}_2 \ldots + c_n\,\mathbf{A}\,\mathbf{X}_n \qquad (5.2)$$

However, the vectors $\mathbf{X}_1$, $\mathbf{X}_2 \ldots \mathbf{X}_n$ are all eigenvectors and so

$$\mathbf{A}\,\mathbf{X}_i = \lambda_i\mathbf{X}_i \qquad (5.3)$$

which means that equation (5.2) becomes

$$\mathbf{V}_1 = c_1\lambda_1\mathbf{X}_1 + c_2\lambda_2\mathbf{X}_2 \ldots c_n\lambda_n\,\mathbf{X}_n \qquad (5.4)$$

Multiplying equation (5.4) by $\mathbf{A}$ gives

$$\mathbf{V}_2 = \mathbf{A}\,\mathbf{V}_1 = \mathbf{A}^2\mathbf{V}_0 = c_1\lambda_1{}^2\mathbf{X}_1 + c_2\lambda_2{}^2\mathbf{X}_2 \ldots c_n\lambda_n{}^2\mathbf{X}_n$$

on making use of equation (5.3).

Repeated multiplication obviously yields

$$\mathbf{V}_{m-1} = \mathbf{A}\,\mathbf{V}_{m-2} = \mathbf{A}^{m-1}\,\mathbf{V}_0 = c_1\lambda_1{}^{m-1}\,\mathbf{X}_1 + c_2\lambda_2{}^{m-1}\mathbf{X}_2 \ldots + c_n\lambda_n^{m-1}\,\mathbf{X}_n \qquad (5.5)$$

$$\mathbf{V}_m = \mathbf{A}\,\mathbf{V}_{m-1} = \mathbf{A}^m\,\mathbf{V}_0 = c_1\lambda_1{}^m\,\mathbf{X}_1 + c_2\lambda_2{}^m\,\mathbf{X}_2 \ldots + c_n\,\lambda_n^m\mathbf{X}_n \qquad (5.6)$$

Without loss of generality we can assume that the largest eigenvalue is in fact λ_1 and if m is taken to be a sufficiently high power the term involving λ_1 will always dominate provided c_1 is non-zero. Hence, if m is taken to be sufficiently large, equations (5.5) and (5.6) are approximately

$$\mathbf{V}_{m-1} = c_1\lambda_1^{m-1}\,\mathbf{X}_1$$

and

$$\mathbf{V}_m = c_1\lambda_1{}^m\mathbf{X}_1$$

An approximation to the largest eigenvector is therefore given by $\dfrac{|\mathbf{V}_{m-1}|}{|\mathbf{V}_m|}$, where $|\mathbf{V}_m|$ means the norm of $\mathbf{V}_m$. Further, the approximation to the eigenvector is given by $\mathbf{V}_m$, suitably normalized so that large numbers are not involved.

For many physical problems it is the largest eigenvector that is important and so this method is useful.

Example 5.1: Find the largest eigenvalue and the corresponding eigenvector of $\mathbf{A}$ where

$$\mathbf{A} = \begin{pmatrix} 5 & 2 \\ 3 & 4 \end{pmatrix}$$

Take as an arbitrary initial vector $\mathbf{V}_0 = \begin{pmatrix} 2 \\ 1 \end{pmatrix}$, then

$$\mathbf{V}_1 = \mathbf{A}\,\mathbf{V}_0 = \begin{pmatrix} 5 & 2 \\ 3 & 4 \end{pmatrix} \begin{pmatrix} 2 \\ 1 \end{pmatrix} = \begin{pmatrix} 12 \\ 10 \end{pmatrix}$$

$$\mathbf{V}_2 = \mathbf{A}\,\mathbf{V}_1 = \begin{pmatrix} 5 & 2 \\ 3 & 4 \end{pmatrix} \begin{pmatrix} 12 \\ 10 \end{pmatrix} = \begin{pmatrix} 80 \\ 76 \end{pmatrix}$$

$$\mathbf{V}_3 = \mathbf{A}\,\mathbf{V}_2 = \begin{pmatrix} 5 & 2 \\ 3 & 4 \end{pmatrix} \begin{pmatrix} 80 \\ 76 \end{pmatrix} = \begin{pmatrix} 552 \\ 544 \end{pmatrix}$$

$$\mathbf{V}_4 = \mathbf{A}\,\mathbf{V}_3 = \begin{pmatrix} 5 & 2 \\ 3 & 4 \end{pmatrix} \begin{pmatrix} 552 \\ 544 \end{pmatrix} = \begin{pmatrix} 3848 \\ 3832 \end{pmatrix}$$

Hence, an estimate of the eigenvector is given by $\begin{pmatrix} 3848 \\ 3832 \end{pmatrix}$ or $\begin{pmatrix} 1{\cdot}0012 \\ 1{\cdot}0000 \end{pmatrix}$, while for the largest eigenvalue improving approximations are given by

$$\frac{|\mathbf{V}_1|}{|\mathbf{V}_0|} = \sqrt{\frac{244}{5}} = \sqrt{488} = 6{\cdot}9857, \quad \frac{|\mathbf{V}_2|}{|\mathbf{V}_1|} = \sqrt{\frac{12176}{244}} = \sqrt{49{\cdot}902}$$

$$= 7{\cdot}0641$$

$$\frac{|\mathbf{V}_3|}{|\mathbf{V}_2|} = \sqrt{\frac{600640}{12176}} = \sqrt{49{\cdot}330} = 7{\cdot}0235$$

and

$$\frac{|\mathbf{V}_4|}{|\mathbf{V}_3|} = \sqrt{\frac{29491322}{600540}} = \sqrt{49{\cdot}100} = 7{\cdot}0071$$

It would clearly be better, especially when the calculations are carried out on a computer, if the normalization procedure was carried out when each successive vector is calculated, noting of course the scalar by which division occurs. The above calculation therefore becomes

$$\mathbf{V}_0 = \begin{pmatrix}2\\1\end{pmatrix} = \sqrt{5}\begin{pmatrix}2/\sqrt{5}\\1/\sqrt{5}\end{pmatrix} = 2.2361\begin{pmatrix}\cdot 8944\\ \cdot 4472\end{pmatrix} = a_0\begin{pmatrix}\cdot 8944\\ \cdot 4472\end{pmatrix},$$

where $a_0 = 2:2361$

$$\mathbf{V}_1 = \mathbf{A}\,\mathbf{V}_0 = a_0\begin{pmatrix}5 & 2\\3 & 4\end{pmatrix}\begin{pmatrix}\cdot 8944\\ \cdot 4472\end{pmatrix} = a_0\begin{pmatrix}5\cdot 3664\\4\cdot 4720\end{pmatrix}$$

$$= a_0(a_1)\begin{pmatrix}\cdot 76820\\ \cdot 64016\end{pmatrix} \text{ where } a_1 = 6\cdot 9857$$

$$\mathbf{V}_2 = \mathbf{A}\,\mathbf{V}_1 = a_0 a_1\begin{pmatrix}5 & 2\\3 & 4\end{pmatrix}\begin{pmatrix}\cdot 76820\\ \cdot 64016\end{pmatrix} = a_0 a_1\begin{pmatrix}5\cdot 1213\\4\cdot 8652\end{pmatrix}$$

$$= a_0 a_1 a_2\begin{pmatrix}\cdot 72500\\ \cdot 68836\end{pmatrix}, \text{ where } a_2 = 7\cdot 0639$$

$$\mathbf{V}_3 = \mathbf{A}\,\mathbf{V}_2 = a_0 a_1 a_2\begin{pmatrix}5 & 2\\3 & 4\end{pmatrix}\begin{pmatrix}\cdot 72500\\ \cdot 68836\end{pmatrix} = a_0 a_1 a_2\begin{pmatrix}5\cdot 0017\\4\cdot 9284\end{pmatrix}$$

$$= a_0 a_1 a_2 a_3\begin{pmatrix}\cdot 71231\\ \cdot 70187\end{pmatrix}, \text{ where } a_3 = 7\cdot 0218$$

$$\mathbf{V}_4 = \mathbf{A}\,\mathbf{V}_2 = a_0 a_1 a_2 a_3\begin{pmatrix}5 & 2\\3 & 4\end{pmatrix}\begin{pmatrix}\cdot 71231\\ \cdot 70181\end{pmatrix}$$

$$= a_0 a_1 a_2 a_3\begin{pmatrix}4\cdot 9653\\4\cdot 9444\end{pmatrix} = a_0 a_1 a_2 a_3 a_4\begin{pmatrix}\cdot 70961\\ \cdot 70562\end{pmatrix},$$

where $a_4 = 7\cdot 0071$

a_0, a_1, a_2, a_3 and a_4 now give the improving approximations to the eigenvalues while the last vector in each row is an estimate of the eigenvector. Note that the correct solution (by solving the characteristic equation) is an eigenvalue of 7 and a normalized eigenvector of $\begin{pmatrix}\cdot 70711\\ \cdot 70711\end{pmatrix}$. The slight differences between the two methods given above arise from the use of five significant figures through the latter calculation.

5.3. The breakdown of the above method

In view of the existence of the term λ_1^m in equation (5.6) the term $c_1\lambda_1^m\mathbf{X}_1$ will always dominate when m is slected to be large enough, provided c_1 is non-zero. However, should c_1 be zero, this term can never dominate and in this event the method will converge onto the largest eigenvalue that has a non-zero coefficient c_1 rather than onto the largest eigenvalue. In practice for a non-symmetric matrix there exists only one way in which c_1

may be zero and that is when the arbitrary vector has inadvertently been selected to be parallel to one of the other eigenvectors. If this is the case then the coefficient corresponding to this eigenvector will be the only non-zero coefficient and in theory convergence occurs in one step. Should this occur, another arbitrary vector has to be selected and the process repeated to find the largest eigenvalue.

Example 5.2: Find the largest eigenvalue of $\mathbf{A} = \begin{pmatrix} 5 & 2 \\ 3 & 4 \end{pmatrix}$.

Select as arbitrary vector $\mathbf{V} = \begin{pmatrix} -2 \\ 3 \end{pmatrix}$, then

$\mathbf{V}_1 = \mathbf{A}\,\mathbf{V}_0 = \begin{pmatrix} 5 & 2 \\ 3 & 4 \end{pmatrix} \begin{pmatrix} -2 \\ 3 \end{pmatrix} = \begin{pmatrix} -4 \\ 6 \end{pmatrix} = 2\begin{pmatrix} -2 \\ 3 \end{pmatrix}$, giving as an eigenvector $\begin{pmatrix} -2 \\ 3 \end{pmatrix}$ and eigenvalue 2. As this is the same matrix as that used in example 5.1, the largest eigenvalue has not been found and another arbitrary vector, for instance, the one already used in the last example, has to be selected.

5.4. The Rayleigh--Schwartz method for a symmetric matrix

If the matrix $\mathbf{A}$ is symmetric, the following method converges more rapidly and is also capable of modification so that the second largest eigenvalue can be determined. One special property of a symmetric matrix is that the eigenvectors form a mutually orthogonal set that can be normalized. If now

$$\mathbf{V}_0 = c_1\mathbf{X}_1 + c_2\mathbf{X}_2 \ldots + c_n\mathbf{X}_n$$

then $\mathbf{X}_i$ is orthogonal to all other vectors $\mathbf{X}_j$ and $\mathbf{X}_i^T\mathbf{X}_i = 1$. Hence

$$\mathbf{V}_0^T\,\mathbf{V}_0 = c_1^2\mathbf{X}_1^T\mathbf{X}_1 + c_2^2\mathbf{X}_2^T\mathbf{X}_2 \ldots + c_n^2\mathbf{X}_n^T\mathbf{X}_n \quad (5.7)$$

In a similar way to that in section 5.2

$$\mathbf{V}_m = \mathbf{A}^m\,\mathbf{V}_0 = c_1\lambda_1^m\mathbf{X}_1 + c_2\lambda_2^m\mathbf{X}_2 \ldots + c_n\lambda_n^m\mathbf{X}_n,$$

while now

$$\mathbf{V}_m^T\mathbf{V}_m = c_1^2\lambda_1^{2m}\mathbf{X}_1^T\mathbf{X}_1 + c_2^2\lambda_2^{2m}\mathbf{X}_2^T\mathbf{X}_2 \ldots + c_n^2\lambda_n^{2m}\mathbf{X}_n^T\mathbf{X}_n$$

$$= c_1^2\lambda_1^2 + c_2^2\lambda_2^2 \ldots + c_n^2\lambda_n^2 \quad (5.8)$$

If again λ_1 is the largest eigenvector, the term $c_1^2 \lambda_1^{2m} \mathbf{X}_1^T \mathbf{X}_1$ will dominate all other terms much more rapidly than the corresponding term did in the previous method.

If we denote by $\mu_{mm-1}, \dfrac{\mathbf{V}_m^T \mathbf{V}_{m-1}}{\mathbf{V}_{m-1}^T \mathbf{V}_{m-1}}$ then μ_{mm-1} gives a

sequence of improving approximations for the largest eigenvalue while again $\mathbf{V}_m$ is an approximation for the corresponding eigenvector.

Example 5.3: Find the largest eigenvalue and corresponding eigenvector of the matrix $\mathbf{A}$, where

$$\mathbf{A} = \begin{pmatrix} 2 & 1 \\ 1 & 2 \end{pmatrix}$$

Choose as an arbitrary vector $\mathbf{V}_0 = \begin{pmatrix} 2 \\ 1 \end{pmatrix}$, say, then

$$\mathbf{V}_0 = \mathbf{A}\,\mathbf{V} = \begin{pmatrix} 2 & 1 \\ 1 & 2 \end{pmatrix} \begin{pmatrix} 2 \\ 1 \end{pmatrix} = \begin{pmatrix} 5 \\ 4 \end{pmatrix}$$

$$\mathbf{V}_1 = \mathbf{A}\,\mathbf{V}_1 = \begin{pmatrix} 2 & 1 \\ 1 & 2 \end{pmatrix} \begin{pmatrix} 5 \\ 4 \end{pmatrix} = \begin{pmatrix} 14 \\ 13 \end{pmatrix}$$

$$\mathbf{V}_3 = \mathbf{A}\,\mathbf{V}_2 = \begin{pmatrix} 2 & 1 \\ 1 & 2 \end{pmatrix} \begin{pmatrix} 14 \\ 13 \end{pmatrix} = \begin{pmatrix} 41 \\ 40 \end{pmatrix}$$

$$\mathbf{V}_4 = \mathbf{A}\,\mathbf{V}_3 = \begin{pmatrix} 2 & 1 \\ 1 & 2 \end{pmatrix} \begin{pmatrix} 41 \\ 40 \end{pmatrix} = \begin{pmatrix} 122 \\ 121 \end{pmatrix}$$

Note that as it is not necessary to-divide the magnitude of two vectors in this method, it is no advantage to normalize the vectors though in practice this may be done if desired. An approximation to the eigenvector is therefore $\begin{pmatrix} 122 \\ 121 \end{pmatrix}$ or $\begin{pmatrix} 1{\cdot}00826 \\ 1{\cdot}00000 \end{pmatrix}$ on removing a factor of 121. From these vectors it is easy to calculate the following,

$$\mathbf{V}_0^T\mathbf{V}_0 = 5, \quad \mathbf{V}_1^T\mathbf{V}_0 = 14, \quad \mathbf{V}_1^T\mathbf{V}_1 = 41,$$

$$\mathbf{V}_2^T\mathbf{V}_1 = 122, \quad \mathbf{V}_2^T\mathbf{V}_2 = 365, \quad \mathbf{V}_3^T\mathbf{V}_2 = 1094,$$

$$\mathbf{V}_3^T\mathbf{V}_3 = 3281, \quad \mathbf{V}_4^T\mathbf{V}_3 = 9842, \quad \mathbf{V}_4^T\mathbf{V}_4 = 29525$$

An improving sequence of values for the largest eigenvalue is now given by

$$\mu_{10} = \frac{\mathbf{V}_1^T\mathbf{V}_0}{\mathbf{V}_0^T\mathbf{V}_0} = \frac{14}{5} = 2{\cdot}8, \quad \mu_{11} = \frac{\mathbf{V}_1^T\mathbf{V}_1}{\mathbf{V}_0^T\mathbf{V}_1} = \frac{41}{14} = 2{\cdot}928511,$$

$$\mu_{21} = \frac{\mathbf{V}_2^T\mathbf{V}_1}{\mathbf{V}_1^T\mathbf{V}_1} = \frac{122}{41} = 2{\cdot}975610, \quad \mu_{22} = \frac{\mathbf{V}_2^T\mathbf{V}_2}{\mathbf{V}_2^T\mathbf{V}_1} = \frac{365}{122}$$

$$= 2{\cdot}991803,$$

$$\mu_{32} = \frac{\mathbf{V}_3^T\mathbf{V}_2}{\mathbf{V}_2^T\mathbf{V}_2} = \frac{1094}{365} = 2{\cdot}997260, \quad \mu_{33} = \frac{\mathbf{V}_3^T\mathbf{V}_3}{\mathbf{V}_3^T\mathbf{V}_2} = \frac{3281}{1094}$$

$$= 2{\cdot}999086,$$

$$\mu_{43} = \frac{\mathbf{V}_4^T\mathbf{V}_3}{\mathbf{V}_3^T\mathbf{V}_3} = \frac{9842}{3281} = 2{\cdot}999695, \quad \mu_{44} = \frac{\mathbf{V}_4^T\mathbf{V}_4}{\mathbf{V}_4^T\mathbf{V}_3} = \frac{29525}{9842}$$

$$= 2{\cdot}999898$$

This last approximation is in fact a very good approximation to the correct value (found to be 3 from the characteristic equation).

Another advantage of the Rayleigh–Schwartz method is that it allows a determination of the second largest eigenvalue. For convenience, and without loss of generality, call the largest eigenvalue λ_1 again, while the second largest eigenvalue is called λ_2. Equations (5.5) and (5.6) now give

$$\mathbf{V}_{m-1} = c_1\lambda_1^{m-1}\mathbf{X}_1 + c_2\lambda_2^{m-1}\mathbf{X}_2 + \ldots.$$

and

$$\mathbf{V}_m = c_1\lambda_1^{m}\mathbf{X}_1 + c_2\lambda_2^{m}\mathbf{X}_2 + \ldots$$

so that

$$\mathbf{V}_m^T\mathbf{V}_{m-1} = c_1^2\lambda_1^{2m-1} + c_2^2\lambda_2^{2m-1} + \ldots$$

and

$$\mathbf{V}_m^T\mathbf{V}_m = c_1^2\lambda_1^{2m} + c_2^2\lambda_2^{2m} + \ldots$$

Hence,

$$\mu_{mm} = \frac{\mathbf{V}_m^T \mathbf{V}_m}{\mathbf{V}_m^T \mathbf{V}_{m-1}} = \frac{c_1^2 \lambda_1^{2m} + c_2^2 \lambda_2^{2m}}{c_1^2 \lambda_1^{2m-1} + c_2^2 \lambda_2^{2m-1}}$$

$$= \lambda_1 + \frac{c_2^2 \lambda_2^{2m} - c_2^2 \lambda_1 \lambda_2^{2m-1}}{c_1^2 \lambda_1^{2m-1} + c_2^2 \lambda_2^{2m-1}}$$

$$= \lambda_1 - \frac{\lambda_1 \lambda_2^{2m-1} \left(1 - \frac{\lambda_2}{\lambda_1}\right) c_2^2}{c_1^2 \lambda_1^{2m-1} + c_2^2 \lambda_2^{2m-1}} \tag{5.9}$$

Therefore, for large m, as $\lambda_1^m >> \lambda_2^m$, equation (5.9) gives

$$\mu_{mm} = \lambda_1 - \lambda_1 \left(\frac{\lambda_2}{\lambda_1}\right)^{2m-1} \left(1 - \frac{\lambda_2}{\lambda_1}\right) \frac{c_2^2}{c_1^2} \tag{5.10}$$

Performing simple algebraic manipulation gives

$$\mu_{mm} - \mu_{m\,m-1} = \lambda_1 \left(\frac{\lambda_2}{\lambda_1}\right)^{2m-2} \left(1 - \frac{\lambda_2}{\lambda_1}\right) \frac{c_2^2}{c_1^2} \left(1 - \frac{\lambda_2}{\lambda_1}\right)$$

and (5.11)

$$\mu_{m\,m-1} - \mu_{m-1\,m-2} = \lambda_1 \left(\frac{\lambda_2}{\lambda_1}\right)^{2m-3} \left(1 - \frac{\lambda_2}{\lambda_1}\right) \frac{c_2^2}{c_1^2} \left(1 - \frac{\lambda_2}{\lambda_1}\right)$$

(5.12)

Dividing equations (5.11) and (5.12) therefore enables a determination of λ_2 to be made as

$$\frac{\mu_{m\,m} - \mu_{m\,m-1}}{\mu_{m\,m-1} - \mu_{m-1\,m-1}} = \frac{\lambda_2}{\lambda_1} \tag{5.13}$$

and λ_1 is already known.

Example 5.4: Find the second largest eigenvalue of the matrix **A** where

$$\mathbf{A} = \begin{pmatrix} 2 & 1 \\ 1 & 2 \end{pmatrix}$$

From example 5·3, $\mu_{10}, \ldots \mu_{44}$ are all known, hence improving estimates for the ratio of the two eigenvalues are given by

$$\frac{\mu_{21} - \mu_{11}}{\mu_{11} - \mu_{10}} = \frac{2{\cdot}975610 - 2{\cdot}928571}{2{\cdot}928571 - 2{\cdot}8} = \frac{0{\cdot}047039}{0{\cdot}128571} = 0{\cdot}36586,$$

$$\frac{\mu_{22} - \mu_{21}}{\mu_{21} - \mu_{11}} = \frac{2{\cdot}991803 - 2{\cdot}975610}{2{\cdot}975610 - 2{\cdot}928571} = \frac{0{\cdot}016193}{0{\cdot}047039} = 0{\cdot}34425,$$

$$\frac{\mu_{32} - \mu_{22}}{\mu_{22} - \mu_{21}} = \frac{2{\cdot}997260 - 2{\cdot}991803}{2{\cdot}991803 - 2{\cdot}975610} = \frac{0{\cdot}005457}{0{\cdot}016193} = 0{\cdot}33700,$$

$$\frac{\mu_{33} - \mu_{32}}{\mu_{32} - \mu_{22}} = \frac{2{\cdot}999086 - 2{\cdot}997260}{2{\cdot}997260 - 2{\cdot}991803} = \frac{0{\cdot}001826}{0{\cdot}005457} = 0{\cdot}33462,$$

$$\frac{\mu_{43} - \mu_{33}}{\mu_{33} - \mu_{32}} = \frac{2{\cdot}999695 - 2{\cdot}999086}{2{\cdot}999086 - 2{\cdot}997260} = \frac{0{\cdot}000609}{0{\cdot}001826} = 0{\cdot}33352,$$

$$\frac{\mu_{44} - \mu_{43}}{\mu_{43} - \mu_{33}} = \frac{2{\cdot}999898 - 2{\cdot}999695}{2{\cdot}999695 - 2{\cdot}999086} = \frac{0{\cdot}000203}{0{\cdot}000609} = 0{\cdot}33333$$

Thus,

$$\lambda_2 = 0{\cdot}33333\lambda_1 = 0{\cdot}99996$$

as the best estimate of the second eigenvalue to the accuracy that we have worked (the true value being 1).

5.5. The breakdown of the Rayleigh–Schwartz method

As for the previous method, the Rayleigh–Schwartz method will converge to the largest eigenvalue if a high enough power is considered, provided that c_1 is not zero, while again if c_1 is zero, it will always give another eigenvalue. As the matrix is now symmetric, the eigenvectors form a mutually orthogonal set and so there are two ways in which c_1 can be zero. First, as for the last method, c_1 will be zero if the arbitrary vector is accidentally selected to be parallel to one of the other vectors, in which case the method converges in one step onto the corresponding eigenvalue. Secondly, c_1 will be zero if the arbitrary vector is inadvertently taken to be orthogonal to the eigenvector corresponding to the largest eigenvalue. If this occurs the method gives the second largest eigenvalue. In fact this second eventuality is unlikely to occur in practice as the small round-off errors will make c_1 very small rather than zero, in which case convergence, though very slow, will still be onto the largest eigenvalue.

5.6. A useful elimination procedure

If λ_1 is a known eigenvalue of the matrix $\mathbf{A}$ and $\mathbf{X}_i$ is the corresponding eigenvector, then

$$\mathbf{A}\,\mathbf{X}_i = \lambda_i\,\mathbf{X}_i$$

Hence, if μ is any scalar

$$(\mathbf{A} - \mu\mathbf{I})\,\mathbf{X}_i = \mathbf{A}\,\mathbf{X}_i - \mu\,\mathbf{X}_i = (\lambda_i - \mu)\,\mathbf{X}_i \qquad (5.14)$$

or

$$\mathbf{B}\,\mathbf{X}_i = (\lambda_i - \mu)\,\mathbf{X}_i$$

where

$$\mathbf{B} = (\mathbf{A} - \mu\mathbf{I}) \qquad (5.15)$$

Equation (5.14) shows that if λ_i is any eigenvalue of $\mathbf{A}$, then $(\lambda_i - \mu)$ is an eigenvalue of the new matrix $\mathbf{B}$ defined by equation (5.15). The corresponding eigenvector is the same for both matrices. This result can be of use in three ways as follows:

(a) If one of the eigenvalues, λ_s say, has been determined by some method or is known, then with

$$\mathbf{B} = \mathbf{A} - \lambda_s\mathbf{I}$$

either of the methods described above can be used to evaluate the eigenvalues of $\mathbf{B}$. Since λ_s is one of the eigenvalues of $\mathbf{A}$, $\lambda_s - \lambda_s = 0$ is an eigenvalue of $\mathbf{B}$. The convergence may now be more rapid as there exists one term fewer that has to be dominated by $c_1\,\lambda_1^m$

Example 5.5: Find the largest eigenvalue of $\mathbf{A}$ where

$$\mathbf{A} = \begin{pmatrix} 2 & 1 \\ 1 & 2 \end{pmatrix}$$

given that one eigenvalue is 1.

Construct the matrix $\mathbf{B} = \mathbf{A} - \mathbf{I} = \begin{pmatrix} 1 & 1 \\ 1 & 1 \end{pmatrix}$.

Selecting as arbitrary vector $\mathbf{V}_0 = \begin{pmatrix} 2 \\ 1 \end{pmatrix}$ again, we have

$$\mathbf{V}_1 = \mathbf{B}\,\mathbf{V}_0 = \begin{pmatrix} 1 & 1 \\ 1 & 1 \end{pmatrix}\begin{pmatrix} 2 \\ 1 \end{pmatrix} = \begin{pmatrix} 3 \\ 3 \end{pmatrix}$$

$$\mathbf{V}_2 = \mathbf{B}\,\mathbf{V}_1 = \begin{pmatrix} 1 & 1 \\ 1 & 1 \end{pmatrix}\begin{pmatrix} 3 \\ 3 \end{pmatrix} = \begin{pmatrix} 6 \\ 6 \end{pmatrix} = 2\begin{pmatrix} 3 \\ 3 \end{pmatrix}$$

Hence, the largest eigenvalue of $\mathbf{B}$ is 2. The eigenvalue of $\mathbf{A}$ is given by $\lambda - 1 = 2$, that is 3. Note that the process is very much more rapid than that used in example 5.3.

(b) The above method may also be used to evaluate smaller eigenvalues when the largest has been evaluated simply by selecting λ_s to be the largest eigenvector.

Example 5.6: Find the second eigenvalue of the matrix $\mathbf{A}$ where

$$\mathbf{A} = \begin{pmatrix} 2 & 1 \\ 1 & 2 \end{pmatrix}$$

given that the largest eigenvalue is 3.

Let $\mathbf{B} = \mathbf{A} - 3\mathbf{I} = \begin{pmatrix} -1 & 1 \\ 1 & -1 \end{pmatrix}$ and choose $\mathbf{V}_0 = \begin{pmatrix} 2 \\ 1 \end{pmatrix}$

$$\mathbf{V}_1 = \mathbf{B}\,\mathbf{V}_0 = \begin{pmatrix} -1 & 1 \\ 1 & -1 \end{pmatrix}\begin{pmatrix} 2 \\ 1 \end{pmatrix} = \begin{pmatrix} -1 \\ 1 \end{pmatrix}$$

$$\mathbf{V}_2 = \mathbf{B}\,\mathbf{V}_1 = \begin{pmatrix} -1 & 1 \\ 1 & -1 \end{pmatrix}\begin{pmatrix} -1 \\ 1 \end{pmatrix} = \begin{pmatrix} 2 \\ -2 \end{pmatrix} = -2\begin{pmatrix} -1 \\ 1 \end{pmatrix}$$

Hence an eigenvalue of $\mathbf{B}$ is -2. The corresponding eigenvalue of $\mathbf{A}$ is therefore given by $-2 + 3 = 1$.

The method still works even if only an approximation to the largest eigenvalue is known, though convergence may be slower.

Example 5.7: Find the second eigenvalue of the matrix $\mathbf{A}$ where

$$\mathbf{A} = \begin{pmatrix} 2 & 1 \\ 1 & 2 \end{pmatrix}$$

given that the largest has been found to be 2·9900.

Construct $\mathbf{B} = \mathbf{A} - 2{\cdot}9900\mathbf{I} = \begin{pmatrix} -0{\cdot}9900 & 1{\cdot}0000 \\ 1{\cdot}0000 & -0{\cdot}9900 \end{pmatrix}$ and let

$$\mathbf{V}_0 = \begin{pmatrix} 2 \\ 1 \end{pmatrix}$$

$$\mathbf{V}_1 = \mathbf{B}\,\mathbf{V}_0 = \begin{pmatrix} -0{\cdot}9800 \\ 1{\cdot}0100 \end{pmatrix}, \quad \mathbf{V}_2 = \mathbf{B}\,\mathbf{V}_1 = \begin{pmatrix} 1{\cdot}9802 \\ -1{\cdot}9799 \end{pmatrix},$$

$$\mathbf{V}_3 = \mathbf{B}\,\mathbf{V}_2 = \begin{pmatrix} -3{\cdot}9403 \\ 3{\cdot}9403 \end{pmatrix}, \quad \mathbf{V}_4 = \mathbf{B}\,\mathbf{V}_3 = \begin{pmatrix} 7{\cdot}8412 \\ -7{\cdot}8412 \end{pmatrix}$$

Hence, the eigenvector is $\begin{pmatrix} 1 \\ -1 \end{pmatrix}$ and the eigenvalue is

$$\frac{-7{\cdot}8412}{3{\cdot}9403} = -1{\cdot}9900.$$

The eigenvalue of $\mathbf{A}$ is therefore $2{\cdot}9900 - 1{\cdot}9900 = 1{\cdot}000$.

(c) The procedure may be used to speed the rate of convergence if the rough magnitude of the eigenvalues is known. Suppose has n eigenvalues, $\lambda_1 \ldots \lambda_n$, say, being all positive and let them be written in descending order. Define

$$\mathbf{B} = \mathbf{A} - \lambda_n \mathbf{I},$$

then the eigenvalues of $\mathbf{B}$ are

$$\lambda_1 - \lambda_n,\ \lambda_2 - \lambda_n,\ \ldots 0$$

But

$$\frac{\lambda_1 - \lambda_n}{\lambda_2 - \lambda_n} > \frac{\lambda_1}{\lambda_2}$$

so that $(\lambda_1 - \lambda_n)$ has to be raised to a smaller power to dominate the other eigenvalues than λ_1 has to. Hence convergence will be more rapid.

Example 5.8: The matrix $\mathbf{A} = \begin{pmatrix} 5 & 2 \\ 3 & 4 \end{pmatrix}$ of example 5·1 has been shown to have one eigenvalue of about 7 and so the second eigenvalue is about 2 (sum of roots = 9). Recalculate these eigenvalues using the above modification.

Estimate the smallest eigenvalue as 1.5 and construct

$$\mathbf{B} = \mathbf{A} - 1.5\mathbf{I} = \begin{pmatrix} 3{\cdot}5 & 2 \\ 3 & 2{\cdot}5 \end{pmatrix}$$

Select $\mathbf{V}_0 = \begin{pmatrix} 2 \\ 1 \end{pmatrix}$ again

$$\mathbf{V}_1 = \begin{pmatrix} 3{\cdot}5 & 2 \\ 3 & 2{\cdot}5 \end{pmatrix}\begin{pmatrix} 2 \\ 1 \end{pmatrix} = \begin{pmatrix} 9 \\ 8{\cdot}5 \end{pmatrix}$$

$$\mathbf{V}_2 = \begin{pmatrix} 3{\cdot}5 & 2 \\ 3 & 2{\cdot}5 \end{pmatrix}\begin{pmatrix} 9 \\ 8{\cdot}5 \end{pmatrix} = \begin{pmatrix} 48{\cdot}5 \\ 48{\cdot}25 \end{pmatrix}$$

$$\mathbf{V}_3 = \begin{pmatrix} 3{\cdot}5 & 2 \\ 3 & 2{\cdot}5 \end{pmatrix}\begin{pmatrix} 48{\cdot}5 \\ 48{\cdot}25 \end{pmatrix} = \begin{pmatrix} 266{\cdot}25 \\ 266{\cdot}125 \end{pmatrix}$$

$$= 266{\cdot}125 \begin{pmatrix} 1{\cdot}000497 \\ 1{\cdot}000000 \end{pmatrix}$$

$$\mathbf{V}_4 = \begin{pmatrix} 3{\cdot}5 & 2 \\ 3 & 2.5 \end{pmatrix}\begin{pmatrix} 266{\cdot}25 \\ 266{\cdot}125 \end{pmatrix} = \begin{pmatrix} 1464{\cdot}125 \\ 1464{\cdot}0625 \end{pmatrix}$$

$$= 1464{\cdot}0625 \begin{pmatrix} 1{\cdot}0000427 \\ 1{\cdot}000000 \end{pmatrix}$$

$$\frac{|\mathbf{V}_4|}{|\mathbf{V}_3|} = \frac{1464{\cdot}0625}{266{\cdot}125} \frac{1 + (1{\cdot}0000427)^2}{1 + (1{\cdot}000497)^2} =$$

$$= (5{\cdot}014091)\,({\cdot}999773) = 5{\cdot}50016$$

We see that $\mathbf{V}_3$ gives a better approximation to the eigenvector than $\mathbf{V}_4$ did in example 5.1, while the estimate for the eigenvalue is $5{\cdot}50016 + 1{\cdot}5 = 7{\cdot}00016$ which is also more accurate than the estimate given in example 5.1.

5.7. Evaluating the smaller eigenvalues

For some types of physical problems it is only the smallest eigenvalues that are important. It will be very time-consuming to repeat the procedure of section 5.6 (b) a sufficient number of times for a large matrix and the following method may be preferable.

If λ and $\mathbf{X}$ are an eigenvalue and corresponding eigenvector of a non-singular matrix $\mathbf{A}$, then

$$\mathbf{A}\,\mathbf{X} = \lambda\,\mathbf{X}$$

Thus

$$\lambda^{-1}\,\mathbf{A}^{-1}\,\mathbf{A}\,\mathbf{X} = \lambda^{-1}\,\lambda\,\mathbf{A}^{-1}\,\mathbf{X}$$

or

$$\lambda^{-1}\mathbf{X} = \mathbf{A}^{-1}\,\mathbf{X} \tag{5.16}$$

which shows that λ^{-1} is an eigenvalue of $\mathbf{A}^{-1}$. Hence, if the largest eigenvalue of the matrix $\mathbf{A}^{-1}$ is found, its reciprocal will give the smallest eigenvalue of $\mathbf{A}$. This method will not work for singular matrices, but for singular matrices the smallest eigenvalue is known to be zero.

Example 5.9: Find the smallest eigenvalue of the matrix $\mathbf{A}$ where

$$\mathbf{A} = \begin{pmatrix} 11 & -9 \\ -9 & 11 \end{pmatrix}$$

By the normal methods the inverse matrix can be calculated as

$$\mathbf{A}^{-1} = \frac{1}{40}\begin{pmatrix} 11 & 9 \\ 9 & 11 \end{pmatrix}$$

Select as arbitrary vector $\mathbf{V}_0 = \begin{pmatrix} 1 \\ 0 \end{pmatrix}$, say, and using the Rayleigh–Schwartz method to calculate the eigenvalues we obtain

$$\mathbf{V}_1 = \frac{1}{40}\begin{pmatrix} 11 & 9 \\ 9 & 11 \end{pmatrix}\begin{pmatrix} 1 \\ 0 \end{pmatrix} = \frac{1}{40}\begin{pmatrix} 11 \\ 9 \end{pmatrix}$$

$$\mathbf{V}_2 = \left(\frac{1}{40}\right)^2\begin{pmatrix} 11 & 9 \\ 9 & 11 \end{pmatrix}\begin{pmatrix} 11 \\ 9 \end{pmatrix} = \left(\frac{1}{40}\right)^2\begin{pmatrix} 202 \\ 198 \end{pmatrix}$$

$$\mathbf{V}_3 = \left(\frac{1}{40}\right)^3\begin{pmatrix} 11 & 9 \\ 9 & 11 \end{pmatrix}\begin{pmatrix} 202 \\ 198 \end{pmatrix} = \left(\frac{1}{40}\right)^3\begin{pmatrix} 4004 \\ 3996 \end{pmatrix}$$

while

$$\mathbf{V}_0{}^T\mathbf{V}_0 = 1,\ \mathbf{V}_1{}^T\mathbf{V}_0 = \frac{11}{40},\ \mathbf{V}_1{}^T\mathbf{V}_1 = \frac{202}{(40)^2},\ \mathbf{V}_2\ \mathbf{V}_1 = \frac{4004}{(40)^3},$$

$$\mathbf{V}_2{}^T\mathbf{V}_2 = \frac{80008}{(40)^4},\ \mathbf{V}_3{}^T\mathbf{V}_2 = \frac{1600016}{(40)^5},\ \mathbf{V}_3{}^T\mathbf{V}_1 = \frac{32000032}{(40)^6}$$

Improving estimates for the eigenvalue of $\mathbf{A}^{-1}$ are therefore given by (leaving out the first few terms)

$$\mu_{22} = \frac{\mathbf{V}_2{}^T\mathbf{V}_2}{\mathbf{V}_2{}^T\mathbf{V}_1} = \frac{80008}{40.4004} = 0{\cdot}499550$$

$$\mu_{32} = \frac{\mathbf{V}_3{}^T\mathbf{V}_2}{\mathbf{V}_2{}^T\mathbf{V}_2} = \frac{1600016}{40.80008} = 0{\cdot}499955$$

and

$$\mu_{33} = \frac{32000032}{40{\cdot}1600016} = 0{\cdot}4999955$$

The corresponding eigenvalue of $\mathbf{A}$ is the reciprocal of that of $\mathbf{A}^{-1}$, that is $\dfrac{1}{0{\cdot}4999955} = 2{\cdot}000018.$

5.8. Jacobi methods

If all the eigenvalues of a symmetric matrix are required, the following method, which is based on geometric intuition, is very helpful. In section 4.6 it was shown that a symmetric matrix of order 2 could always be reduced to a diagonal matrix by means of an orthogonal matrix of the form

$$\begin{pmatrix} \cos\theta & \sin\theta \\ -\sin\theta & \cos\theta \end{pmatrix}$$

where θ was interpreted as the angle through which the reference axis had to be rotated. This being the case, the eigenvalues of a matrix of order 2 can be found as follows.

Let $\begin{pmatrix} a & b \\ b & c \end{pmatrix}$ be the matrix, then

$$\begin{pmatrix} \cos\theta & -\sin\theta \\ \sin\theta & \cos\theta \end{pmatrix} \begin{pmatrix} a & b \\ b & c \end{pmatrix} \begin{pmatrix} \cos\theta & \sin\theta \\ -\sin\theta & \cos\theta \end{pmatrix} = \begin{pmatrix} \lambda_1 & 0 \\ 0 & \lambda_2 \end{pmatrix} \tag{5.17}$$

provided θ is suitably chosen

Performing the multiplication implied in equation (5.17) we obtain

$$\begin{pmatrix} f(\theta) & (a-c)\sin\theta\cos\theta + b(\cos^2\theta - \sin^2\theta) \\ (a-c)\sin\theta\cos\theta + b(\cos^2\theta - \sin^2\theta) & f(\theta) \end{pmatrix}$$

$$= \begin{pmatrix} \lambda_1 & 0 \\ 0 & \lambda_2 \end{pmatrix}$$

where $f(\theta)$ and $g(\theta)$ are some functions of θ that can be determined. In order to satisfy this equality, we must have

$$(c-a)\sin\theta\cos\theta = b(\cos^2\theta - \sin^2\theta)$$

or

$$\tan 2\theta = \frac{2b}{c-a} \tag{5.18}$$

When θ has been determined from equation (5.18), evaluation of $f(\theta)$ and $g(\theta)$ from equation (5.17) gives the eigenvalues.

Consider now a larger matrix **A** or order n, with elements $a_{\alpha\alpha}$, $a_{\alpha\beta}$, $a_{\beta\alpha}$ and $a_{\beta\beta}$ occurring in row α column α, row α column β, row β column α and row β column β. Consider the orthogonal matrix $\mathbf{Q}_{-1}$ where all the off-diagonal elements are zero with the exception of $q_{\alpha\beta}$ and $q_{\beta\alpha}$ and where all the diagonal elements are unity with the exception of $q_{\alpha\alpha}$ and $q_{\beta\beta}$. Also let $q_{\alpha\alpha} = q_{\beta\beta} = \cos\theta$, $q_{\alpha\beta} = \sin\theta$ and $q_{\beta\alpha} = -\sin\theta$.

The transformation $\mathbf{Q}_1{}^T \mathbf{A} \mathbf{Q}_1$ will leave all but rows α, β and columns α, β unaltered and if θ is so selected that

$$\tan 2\theta = \frac{2\,a_{\alpha\beta}}{a_{\beta\beta} - a_{\alpha\alpha}}$$

then the elements in row α column β and row β column α will be reduced to zero by the transformation.

The procedure then is to locate the largest pair of off-diagonal elements and calculate the necessary angle θ_1 to reduce these to zero by a transformation $\mathbf{Q}^T{}_1 \mathbf{A} \mathbf{Q}_1$. The process is then repeated locating the largest pair of diagonal elements in $\mathbf{Q}_1^T \mathbf{A} \mathbf{Q}_1$ and reducing them to zero using the necessary value θ_2 in a transformation $\mathbf{Q}_1^T \mathbf{Q}_1^T \mathbf{A} \mathbf{Q}_1 \mathbf{Q}_2$. This will possibly destroy the first of pair of null elements generated but since all the matrices $\mathbf{Q}$ are orthogonal the new elements will be smaller than the original. The process is repeated until all the off-diagonal elements are acceptably close to zero.

Example 5.10: Find the eigenvalues of $\mathbf{A}$ where

$$\mathbf{A} = \begin{pmatrix} 15 & \sqrt{6} & 7 \\ \sqrt{6} & 18 & \sqrt{6} \\ 7 & \sqrt{6} & 15 \end{pmatrix}$$

The largest off-diagonal elements occur in row 1 column 3 and row 3 column 1.

Select therefore θ_1 so that

$$tan\ 2\theta_1 = \frac{2.7}{15\text{–}15}$$

which gives $\theta_1 = \pi/4$. Operating with the matrix $\mathbf{Q}_1$, where

$$\mathbf{Q}_1 = \begin{pmatrix} cos\,\pi/4 & 0 & sin\,\pi/4 \\ 0 & 1 & 0 \\ -sin\,\pi/4 & 0 & cos\,\pi/4 \end{pmatrix} = \frac{1}{\sqrt{2}} \begin{pmatrix} 1 & 0 & 1 \\ 0 & \sqrt{2} & 0 \\ -1 & 0 & 1 \end{pmatrix}$$

we obtain

$$\mathbf{Q}_1^T \mathbf{A} \mathbf{Q} = \frac{1}{2} \begin{pmatrix} 1 & 0 & -1 \\ 0 & \sqrt{2} & 0 \\ 1 & 0 & 1 \end{pmatrix} \begin{pmatrix} 15 & \sqrt{6} & 7 \\ \sqrt{6} & 18 & \sqrt{6} \\ 7 & \sqrt{6} & 15 \end{pmatrix} \begin{pmatrix} 1 & 0 & 1 \\ 0 & \sqrt{2} & 0 \\ -1 & 0 & 1 \end{pmatrix}$$

$$= \begin{pmatrix} 8 & 0 & 0 \\ 0 & 18 & \sqrt{12} \\ 0 & \sqrt{12} & 22 \end{pmatrix}$$

There now exists only one pair of off-diagonal elements; selecting θ_2 so that

$$tan\,2\theta_2 = \frac{2\sqrt{12}}{22\text{–}18} = \sqrt{3},$$

or
$$\theta_2 = \pi/6,$$

and
$$\mathbf{Q}_2 = \begin{pmatrix} 1 & 0 & 0 \\ 0 & \cos\pi/6 & \sin\pi/6 \\ 0 & -\sin\pi/6 & \cos\pi/6 \end{pmatrix} = \begin{pmatrix} 1 & 0 & 0 \\ 0 & \sqrt{3}/2 & \frac{1}{2} \\ 0 & -\frac{1}{2} & 3/2 \end{pmatrix}$$

gives

$$\mathbf{Q}_2{}^T \mathbf{Q}_1{}^T \mathbf{A}\,\mathbf{Q}_1 \mathbf{Q}_2 = \begin{pmatrix} 1 & 0 & 0 \\ 0 & \sqrt{3}/2 & -\frac{1}{2} \\ 0 & \frac{1}{2} & \sqrt{3}/2 \end{pmatrix} \begin{pmatrix} 8 & 0 & 0 \\ 0 & 18 & \sqrt{12} \\ 0 & \sqrt{12} & 22 \end{pmatrix} \begin{pmatrix} 1 & 0 & 0 \\ 0 & \sqrt{3}/2 & \frac{1}{2} \\ 0 & -\frac{1}{2} & \sqrt{3}/2 \end{pmatrix}$$

$$= \begin{pmatrix} 8 & 0 & 0 \\ 0 & 16 & 0 \\ 0 & 0 & 24 \end{pmatrix}$$

The eigenvalues are therefore 8, 16 and 24.

EXERCISES 5

1. Find the largest eigenvalue of the matrix **A** where **A** is given by

$$\begin{pmatrix} 4 & 3 \\ 1 & 2 \end{pmatrix}, \begin{pmatrix} 10 & 9 \\ 5 & 6 \end{pmatrix}, \begin{pmatrix} 6 & 5 \\ -4 & 5 \end{pmatrix}$$

2. Find the two largest eigenvalues of **A**, where

$$\mathbf{A} = \begin{pmatrix} 5 & -2 & 0 \\ -2 & 6 & 2 \\ 0 & 2 & 7 \end{pmatrix}$$

using the Rayleigh–Schwartz method.

3. Find the other two eigenvalues of **A**, where

$$\mathbf{A} = \begin{pmatrix} 1 & 0 & -4 \\ 0 & 5 & 4 \\ -4 & 4 & 3 \end{pmatrix}$$

given that one eigenvalue is -3.

4. Find the smallest eigenvalue of $\mathbf{A}$, where

$$\mathbf{A} = \begin{pmatrix} 2 & 1 \\ 1 & 2 \end{pmatrix}$$

5. Use the Jacobi method to find the eigenvalues of

$$\begin{pmatrix} 0 & 0 & 1 \\ 0 & 1 & 0 \\ 1 & 0 & 0 \end{pmatrix}$$

6

SOME MATRIX APPLICATIONS

6.1. Introduction

In this chapter a number of situations are discussed where matrices are used in solving a problem arising in another field. Every field of application has not been covered; instead an attempt has been made to illustrate most of the major properties of matrices by including one typical application. There are two major divisions which can be made, applications where the multiplication of matrices and calculating the inverse matrix or solving linear equations are involved and applications where the evaluations of eigenvalues, mostly to be interpreted as frequencies, are involved.

6.2. Applications of matrix algebra

Use of matrix multiplication

Table 6.1 illustrates the price of five commodities, tea, coffee, sugar, bread and eggs, in six different shops A,B,C,D,E and F.

TABLE 6.1

	Tea	Coffee	Sugar	Bread	Eggs
Shop A	14	30	16	10	22
Shop B	15	31	14	11	22
Shop C	13	29	12	11	23
Shop D	14	31	15	10	21
Shop E	16	30	16	11	21
Shop F	13	29	15	11	23

Four people, John, William, Peter and Nicholas, each purchase some of these commodities, the amounts bought being shown in Table 6.2.

TABLE 6.2

	John	William	Peter	Nicholas
Tea	1	0	2	1
Coffee	0	2	0	1
Sugar	2	1	2	3
Bread	4	5	3	4
Eggs	2	1	3	2

It is required to find the cost to each person of doing all his shopping in each of the five shops. The procedure is fairly obvious: the quality of goods bought has to be multiplied by the cost of that produce and the sum of all such multiplications taken, the resulting cost being shown in Table 6.3.

TABLE 6.3

	John	William	Peter	Nicholas
Shop A	130	148	156	176
Shop B	131	153	157	176
Shop C	127	148	152	168
Shop D	126	148	151	172
Shop E	134	152	160	180
Shop F	129	146	155	173

It is therefore cheaper for John to shop at Shop D, William at Shop F, Peter at Shop D, and for Nicholas to shop at Shop C.

Note, however, that the simplest way of generating Table 6.3 from Tables 6.1 and 6.2 is by regarding all the tables as matrices and taking the product of the first two, so that

$$\begin{pmatrix} 14 & 30 & 16 & 10 & 22 \\ 15 & 31 & 14 & 11 & 22 \\ 13 & 29 & 12 & 11 & 23 \\ 14 & 31 & 15 & 10 & 21 \\ 16 & 30 & 16 & 11 & 21 \\ 13 & 29 & 15 & 10 & 23 \end{pmatrix} \begin{pmatrix} 1 & 0 & 2 & 1 \\ 0 & 2 & 0 & 1 \\ 2 & 1 & 2 & 3 \\ 4 & 5 & 3 & 4 \end{pmatrix}$$

$$= \begin{pmatrix} 130 & 148 & 156 & 176 \\ 131 & 153 & 157 & 176 \\ 127 & 148 & 152 & 168 \\ 126 & 148 & 155 & 172 \\ 134 & 152 & 160 & 180 \\ 129 & 146 & 155 & 173 \end{pmatrix} \qquad (6.1)$$

The advantage of introducing matrices into this type of calculation is that programs already exist in computer libraries for the multiplication of matrices while a program would have to be developed for evaluating Table 6.3 from Tables 6.1 and 6.2.

Problems involving linear equations

(a) As a result of performing an experiment, a set of values y_i have been found corresponding to the values x_i of an independent variable. Inspection suggests that a plausible relationship between x and y is of the form

$$y = a + bx + cx^2 \qquad (6.2)$$

say.

It is required to find the values of the coefficients a, b and c such that the sums of the squares of the differences between the experimental values of y and those predicted by the formula are as small as possible, that is if

$$E = \sum_i (a + bx_i + cx_i^2 - y_i)^2$$

it is required that E is a minimum. The condition for such a minimum is that $\dfrac{\delta E}{\delta a}$, $\dfrac{\delta E}{\delta b}$ and $\dfrac{\delta E}{\delta c}$ are all zero, or

$$\begin{aligned} &\sum_i (a + bx_i + cx_i^2 - y_i) = 0 \\ &\sum_i (a + bx_i + cx_i^2 - y_i)\, x_i = 0 \\ &\sum_i (a + bx_i + cx_i^2 - y_i) x_i^2 = 0 \end{aligned} \qquad (6.3)$$

Denoting the summations arising out of these equations, all of which can be evaluated from the experimental results, in the following way,

$$\sum x_i = S_1,\ \sum x_i^2 = S_2,\ \sum x_i^3 = S_3,\ \sum x_i^4 = S_4$$

$$\sum y_i = P_0,\ \sum x_i y_i = P_1,\ \sum x_i^2 = P_2$$

the above equations become

$$na + bS_1 + cS_2 = P_0$$

$$aS_1 + bS_2 + cS_3 = P_1$$

$$aS_2 + bS_3 + cS_4 = P_2$$

$$\text{or}\quad \begin{pmatrix} n & S_1 & S_2 \\ S_1 & S_2 & S_3 \\ S_2 & S_3 & S_4 \end{pmatrix}\begin{pmatrix} a \\ b \\ c \end{pmatrix} = \begin{pmatrix} P_0 \\ P_1 \\ P_2 \end{pmatrix} \qquad (6.4)$$

a set of equations to determine a, b and c which can easily be solved by any of the standard methods given in Chapter 3.

(b) Any point of a block of elastic material in its unstretched shape has coordinates (x, y, z). If this block is slightly stretched in the x-direction, this point will have new cordinates given (to 1st order) by

$$x' = \lambda_1 x,$$

$$y' = y + ax$$

$$z' = z + bx,$$

where λ_1 is greater than unity and both a and b are very small. This deformation can be written as

$$\begin{pmatrix} x' \\ y' \\ z' \end{pmatrix} = \begin{pmatrix} \lambda_1 & 0 & 0 \\ a & 1 & 0 \\ b & 0 & 1 \end{pmatrix}\begin{pmatrix} x' \\ y \\ z \end{pmatrix} \qquad (6.5)$$

If this block were now to be deformed in the y direction, the new coordinates of the point will be

$$\begin{pmatrix} x'' \\ y'' \\ z'' \end{pmatrix} = \begin{pmatrix} 1 & c & 0 \\ 0 & \lambda_2 & 0 \\ 0 & d & 1 \end{pmatrix}\begin{pmatrix} x' \\ y' \\ z' \end{pmatrix}$$

so that

$$\begin{pmatrix} x'' \\ x'' \\ z'' \end{pmatrix} = \begin{pmatrix} 1 & c & 0 \\ 0 & \lambda_2 & 0 \\ 0 & d & 1 \end{pmatrix} \begin{pmatrix} \lambda_1 & 0 & 0 \\ a & 1 & 0 \\ b & 0 & 1 \end{pmatrix} \begin{pmatrix} x \\ y \\ z \end{pmatrix}$$

$$= \begin{pmatrix} \lambda_1 + ac & c & 0 \\ a\lambda_2 & \lambda_2 & 0 \\ ad + b & d & 1 \end{pmatrix} \begin{pmatrix} x \\ y \\ z \end{pmatrix}$$

Note that this is not the same as the position obtained when a deformation along the y-axis is followed by a deformation along the x-axis. A further deformation along the z-axis results in another matrix and in fact a general small deformation can always be expressed as

$$\begin{pmatrix} x' \\ y' \\ z' \end{pmatrix} = \begin{pmatrix} a_{11} & a_{12} & a_{13} \\ a_{21} & a_{22} & a_{23} \\ a_{31} & a_{32} & a_{33} \end{pmatrix} \begin{pmatrix} x \\ y \\ z \end{pmatrix} \qquad (6.6)$$

In practice one might require the old position of some point, given that its position after some deformation is known, so that the above system of equations has to be solved by any of the standard methods, for example, Gaussian elimination.

(c) A strut of length a is subject to a thrust P and has a variable flexual rigidity B. Taking the x-axis to be along the strut and the y-axis transverse to this, the governing differential equation is

$$B \frac{d^2y}{dx^2} + Py = 0$$

and this requires integrating.

One approach to this problem is to adopt a finite difference scheme. In such an approach the strut is imagined to be divided into n equal parts, each part being of length $h = a/n$.

Let $x_m = mh$, while the value of y at x_m is denoted by y_m and the value of B by B_m.

By using Taylor's theorem we obtain

$$\left(\frac{d^2y}{dx^2}\right)_m = \frac{1}{h^2}(y_{m+1} - 2y_m + y_{m-1}) + 0(h^2)$$

and so the differential equation to be integrated becomes

$$B_m(y_{m+1} - 2y_m + y_{m-1}) + Ph^2y_m = 0$$

Normally, a strut would be supported at both ends, so that

$$y_0 = 0 \text{ and } y_n = 0$$

Writing down the equations for each part of the strut gives

$$m = 1 \quad B_1y_2 + (Ph^2 - 2B_1)y_1 + B_1y_0 = 0$$

$$m = 2 \quad B_2y_3 + (Ph^2 - 2B_2)y_2 + B_2y_1 = 0$$

$$\ldots \quad \ldots \quad \ldots \quad \ldots \quad \ldots$$

$$m = n - 1 \quad B_{n-1}y_n + (Ph^2 - 2B_{n-1})y_{n-1} + B_{n-1}y_{n-2} = 0$$

This set can be written in matrix form as

$$\begin{pmatrix} Ph^2 - 2B_1 & B_1 & 0 & \ldots & 0 \\ B_2 & Ph^2 - 2B_2 & B_2 & \ldots & 2B_2 \\ \ldots & \ldots & \ldots & \ldots & \ldots \\ \ldots & \ldots & \ldots & \ldots & \ldots \\ 0 & \ldots & \ldots & B_{n-1} & Ph^2 - 2B_{n-1} \end{pmatrix} \begin{pmatrix} y_1 \\ y_2 \\ \ldots \\ \ldots \\ y_{n-1} \end{pmatrix} = \mathbf{0}$$

which is a set of homogeneous equations that can be solved in the standard way.

(d) A standard Wheatstone bridge network is shown in diagram 6.1. The problem is to calculate the currents and voltage drops in the branches generated by a source of electro-motive force somewhere in the network, assuming that the resistances of all the branches are given. In the usual application of the Wheatstone bridge the resistances R_1, R_2, R_5 and R_4 are adjusted until there is no current flowing through the resistance R_3. We shall consider the more general problem outlined at the beginning of this paragraph.

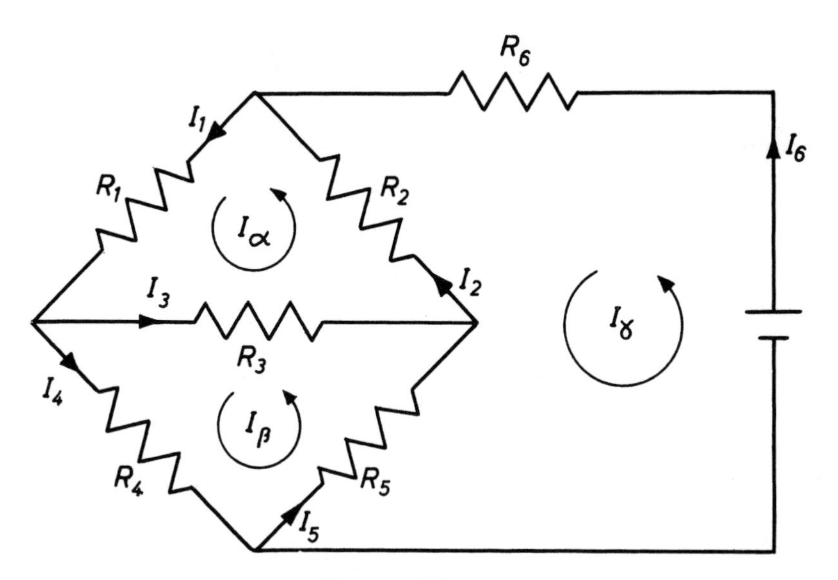

Diagram 6.1

Although this would at first appear to be a six variable problem, using the conservation of current (Kirchhoff's first law) it is reduced to a three variable problem as follows:

$$
\begin{aligned}
I_1 &= I_\alpha \\
I_2 &= I_\alpha - I_\gamma \\
I_3 &= I_\alpha - I_\beta \\
I_4 &= I_\beta \\
I_5 &= I_\beta - I_\gamma \\
I_6 &= I_\gamma
\end{aligned}
$$

where I_α, I_β and I_γ are used to denote the currents in the three closed meshes as shown in diagram 6.1.

These six equations can be rewritten in a matrix form as

$$
\begin{pmatrix} I_1 \\ I_2 \\ I_3 \\ I_4 \\ I_5 \\ I_6 \end{pmatrix} = \begin{pmatrix} 1 & 0 & 0 \\ 1 & 0 & -1 \\ 1 & -1 & 0 \\ 0 & 1 & 0 \\ 0 & 1 & -1 \\ 0 & 0 & 1 \end{pmatrix} \begin{pmatrix} I_\alpha \\ I_\beta \\ I_\gamma \end{pmatrix} \tag{6.7}
$$

or $$\mathbf{I} = \mathbf{A}\,\mathbf{I}', \text{ say}$$

Kirchhoff's second law states that the electromotive force around any closed circuit is zero and applying this to the circuits in question gives

$$\begin{aligned}(R_1 + R_2 + R_3)I_\alpha - R_3 I_\beta + R_2 I_\gamma &= 0\\ -R_3 I_\alpha + (R_3 + R_4 + R_5)I_\beta - R_5 I_\gamma &= 0\\ -R_2 I_\alpha - R_5 I_\beta + (R_2 + R_5 + R_6)I_\gamma &= E\end{aligned} \tag{6.8}$$

This is a set of linear equations of the type that has been considered extensively in Chapter 3 and so I_α, I_β and I_γ can easily be calculated. Using e_i to denote the voltage drop in any branch, these can be calculated, using Ohms's law, from a knowledge of the corresponding currents as

$$e_i = R_i I_i \text{ for all } i \text{ satisfying } 1 \leqslant i \leqslant 5$$

and $$e_6 = R_6 I_6 - E \text{ or } e_6 + E = R_6 I_6$$

Hence, if e is the vector whose elements are the voltage drops, we have

$$\mathbf{e} = \mathbf{diagR}\,\mathbf{I} \tag{6.9}$$

where **diagR** is the diagonal matrix whose elements are the resistances in the branches. But the connection between **I** and the currents I_α, I_β and I_γ is known and so

$$\mathbf{e} = \mathbf{diagR}\,\mathbf{A}\,\mathbf{I}' \tag{6.10}$$

which allows a determination of the voltage drops.

(e) In engineering problems, the displacements of points in a structure, **d**, and the forces applied at those points, **P**, are connected by the stiffness matrix, **K**, such that

$$\mathbf{P} = \mathbf{K}\,\mathbf{d} \tag{6.11}$$

In many situations, only the first few forces, $\mathbf{P}_1$, say, are known but not the corresponding displacements, $\mathbf{d}_1$, while the remaining displacements, $\mathbf{d}_2$, are known but not the forces, $\mathbf{P}_2$. The stiffness matrix is known. This problem is best tackled by partitioning the stiffness matrix so that

$$\mathbf{P}_1 = \mathbf{K}_{11}\mathbf{d}_1 + \mathbf{K}_{12}\mathbf{d}_2$$

and

$$\mathbf{P}_2 = \mathbf{K}_{21}\mathbf{d}_1 + \mathbf{K}_{22}\mathbf{d}_2$$

Then $\mathbf{d}_1 = \mathbf{K}_{11}^{-1}(\mathbf{P}_1 - \mathbf{K}_{12}\mathbf{d}_2)$ and substitution gives

$$\mathbf{P}_2 = \mathbf{K}_{21}\mathbf{K}_{11}^{-1}(\mathbf{P}_1 - \mathbf{K}_{12}\mathbf{d}_2) + \mathbf{K}_{22}\mathbf{d}_2 \qquad (6.12)$$

An example using numerical values that are reasonable, though rounded off to allow easy manipulation, is

$$\begin{pmatrix} 100 \\ 99 \\ P_3 \\ P_4 \end{pmatrix} = \begin{pmatrix} 1000 & 10 & -1000 & -10 \\ 1010 & 2000 & -20 & -1000 \\ -1000 & -20 & 2000 & 10 \\ -10 & -1000 & 10 & 1000 \end{pmatrix} \begin{pmatrix} d_1 \\ d_2 \\ 0{\cdot}2 \\ 0{\cdot}1 \end{pmatrix}$$

Multiplying out gives

$$\begin{pmatrix} 100 \\ 99 \end{pmatrix} = \begin{pmatrix} 1000 & 10 \\ 10 & 2000 \end{pmatrix} \begin{pmatrix} d_1 \\ d_2 \end{pmatrix} - \begin{pmatrix} 1000 & 10 \\ 20 & 1000 \end{pmatrix} \begin{pmatrix} 0{\cdot}2 \\ 0{\cdot}1 \end{pmatrix}$$

$$\begin{pmatrix} 1000 & 10 \\ 10 & 2000 \end{pmatrix} \begin{pmatrix} d_1 \\ d_2 \end{pmatrix} = \begin{pmatrix} 100 \\ 99 \end{pmatrix} + \begin{pmatrix} 201 \\ 104 \end{pmatrix} = \begin{pmatrix} 301 \\ 203 \end{pmatrix}$$

$$\begin{pmatrix} d_1 \\ d_2 \end{pmatrix} = \begin{pmatrix} 1000 & 10 \\ 10 & 2000 \end{pmatrix}^{-1} \begin{pmatrix} 301 \\ 203 \end{pmatrix} = \frac{1}{1{,}999{,}900} \begin{pmatrix} 2000 & -10 \\ -10 & 1000 \end{pmatrix} \begin{pmatrix} 301 \\ 203 \end{pmatrix}$$

$$= \frac{1}{1{,}999{,}900} \begin{pmatrix} 599{,}970 \\ 199{,}990 \end{pmatrix} = \begin{pmatrix} 0{\cdot}3 \\ 0{\cdot}1 \end{pmatrix}$$

and

$$\begin{pmatrix} P_3 \\ P_4 \end{pmatrix} = -\begin{pmatrix} 1000 & 20 \\ 10 & 1000 \end{pmatrix} \begin{pmatrix} 0{\cdot}3 \\ 0{\cdot}1 \end{pmatrix} + \begin{pmatrix} 2000 & 10 \\ 10 & 1000 \end{pmatrix} \begin{pmatrix} 0{\cdot}2 \\ 0{\cdot}1 \end{pmatrix}$$

$$= -\begin{pmatrix} 302 \\ 103 \end{pmatrix} + \begin{pmatrix} 401 \\ 102 \end{pmatrix} = \begin{pmatrix} 99 \\ -1 \end{pmatrix}$$

The unknown displacements therefore were $\begin{pmatrix} 0{\cdot}3 \\ 0{\cdot}1 \end{pmatrix}$, while the unknown forces were $\begin{pmatrix} 99 \\ -1 \end{pmatrix}$

6.3. Use of eigenvalue properties

Solving linear differential equations

Consider a set of differential equations of the form

$$\begin{aligned} \dot{x}_1 &= a_{11}x_1 + a_{12}x_2 \ \dots \ a_{1n}x_n \\ \dot{x}_2 &= a_{21}x_1 + a_{22}x_2 \ \dots \ a_{2n}x_n \\ \dots & \quad \dots \quad \dots \quad \dots \quad \dots \\ \dot{x}_n &= a_{n1}x_1 + a_{n2}x_2 \ \dots \ a_{nn}x_n \end{aligned} \tag{6.13}$$

which can be written in matrix form as

$$\ddot{\mathbf{X}} = \mathbf{A}\,\mathbf{X}$$

Now, if the eigenvalues $\lambda_1, \dots, \lambda_n$ of **A** and its eigenvectors can be found, then the matrix of eigenvectors, **Q**, transforms **A** such that

$$\mathbf{Q}^{-1}\mathbf{A}\,\mathbf{Q} = \Lambda$$

where Λ is the diagonal matrix whose elements are the eigenvalues.

Transforming the variables to

$$\mathbf{X} = \mathbf{Q}\,\mathbf{Y} \tag{6.14}$$

on differentiating, we obtain

$$\dot{\mathbf{X}} = \mathbf{Q}\,\dot{\mathbf{Y}} \tag{6.15}$$

as **Q** is the matrix of eigenvector of **A** and so is time independent. Substituting in the differential equation gives

$$\mathbf{Q}\,\dot{\mathbf{Y}} = \mathbf{A}\,\mathbf{Q}\,\mathbf{Y}$$

or

$$\dot{\mathbf{Y}} = \mathbf{Q}^{-1}\mathbf{A}\,\mathbf{Q}\,\mathbf{Y} = \Lambda\,\mathbf{Y} \tag{6.16}$$

In this last equation, since Λ is a diagonal matrix each unknown occurs in different equations so that

$$\dot{y}_i = \lambda_i y_i \tag{6.17}$$

for every value of i. Each such equation can be integrated to give

$$y_i = b_i e^{\lambda_i t}$$

or

$$\mathbf{Y} = \exp\Lambda t\ \mathbf{B} = \mathbf{C} \tag{6.18}$$

where $\mathbf{B}$ is the vector whose elements are b_i and a usual $\mathbf{exp\Lambda t}$ is the diagonal matrix whose elements are $e^{\lambda_i t}$ while $\mathbf{C}$ is the vector whose elements are $b_i\, e^{\lambda_i t}$. But it is a solution for $\mathbf{X}$ that is required and this is given, after using equation (6.14), as

$$\mathbf{X} = \mathbf{Q}\,\mathbf{Y} = \mathbf{Q}\,\mathbf{exp\Lambda t}\,\mathbf{B} = \mathbf{Q}\,\mathbf{C} \tag{6.19}$$

or

$$\mathbf{X} = \sum_i \mathbf{X}_i\, b_i\, e_i^{\lambda_i t}$$

where $\mathbf{X}_i$ is the eigenvector corresponding to the eigenvalue λ_i, which gives the required solution. There are three major points of interest arising out of the above work.

First, use could have been made of $e^{\mathbf{A}}$ as has been defined by us. The solution to the differential equation

$$\dot{\mathbf{X}} = \mathbf{A}\,\mathbf{X}$$

then becomes

$$\mathbf{X} = e^{\mathbf{A}t}\,\mathbf{B}' \tag{6.20}$$

where $\mathbf{B}'$ is an arbitrary constant vector. However, if $\mathbf{Q}$ again denotes the matrix of eigenvectors of $\mathbf{A}$

$$e^{\mathbf{A}t} = \mathbf{Q}\,\mathbf{exp\Lambda t}\,\mathbf{Q}^{-1}$$

according to the work of Chapter 4. Substituting in equation (6.20) gives

$$\mathbf{X} = \mathbf{Q}\,\mathbf{exp\Lambda t}\,\mathbf{Q}^{-1}\,\mathbf{B}' = \mathbf{Q}\,\mathbf{exp\Lambda t}\,\mathbf{B} \tag{6.21}$$

which agrees with the solution that has already been obtained.

Secondly, since the point of the above method is to generate a set of equations like equation (6.16) where each unknown occurs in a seperate equation, the order of the differential could be higher though the same in all the equations. The only modification would be to the form of the solution from equation (6.18) onwards.

Thirdly, note that differential equations of the form

$$\mathbf{B}\,\dot{\mathbf{X}} = \mathbf{A}\,\mathbf{X} \tag{6.22}$$

where $\mathbf{A}$ and $\mathbf{B}$ are square matrices, can also be solved provided $\mathbf{B}^{-1}$ exists, for then equation (6.22) becomes

$$\dot{\mathbf{X}} = \mathbf{B}^{-1}\mathbf{A}\,\mathbf{X} = \mathbf{C}\,\mathbf{X} \tag{6.23}$$

where $\mathbf{C} = \mathbf{B}^{-1}\mathbf{A}$. Now equation (2.23) is identical to equation (6.14). Note that in fact it is not necessary to calculate $\mathbf{B}^{-1}$, only to verify that it exists, for if λ is an eigenvalue of $\mathbf{C}$ then

$$|\mathbf{C} - \lambda\mathbf{I}| = 0 \text{ or } |\mathbf{B}^{-1}\mathbf{A} - \lambda\mathbf{I}| = 0$$

which implies that $|\mathbf{A} - \lambda\,\mathbf{B}| = 0$ (6.24)

Hence, to obtain the eigenvalues of $\mathbf{C}$, the values of λ satisfying equation (6.24) have to be calculated. Further, if

$$(\mathbf{C} - \lambda\mathbf{I})\,\mathbf{X} = 0$$

then
$$(\mathbf{B}^{-1}\mathbf{A} - \lambda\,\mathbf{I})\mathbf{X} = 0$$

or on multyplying by $\mathbf{B}$

$$(\mathbf{A} - \lambda\,\mathbf{B})\,\mathbf{X} = 0$$

so that the eigenvector is also given by considering the matrix $\mathbf{A} - \lambda\mathbf{B}$.

The application of the above theory of differential equations is widespread. We shall discuss two simple cases here, both of which give an insight to the interpretation of eigenvalues as frequencies of oscillation in some sense.

(a) A string of length $4a$ is taut with its two ends fixed. Three particles, each of mass m, are fixed at distances a, $2a$ and $3a$ from one end of the string. It is required to discuss the motion of these particles after they have been displaced slightly from their equilibrium position. The situation is illustrated in diagram 6.2.

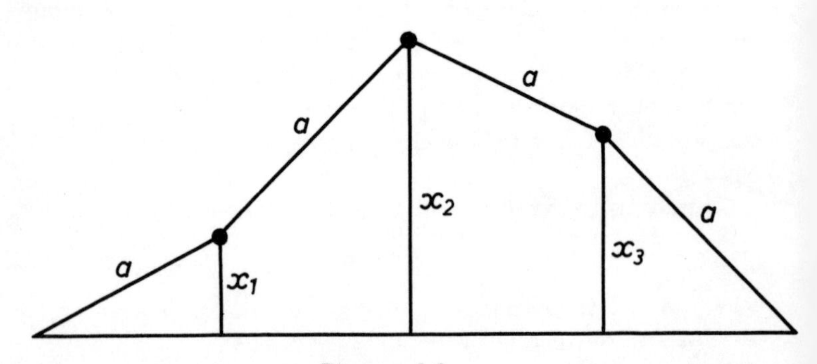

Diagram 6.2

Let the displacements of the particles be x_1, x_2 and x_3, while the tension in the string is T. The equations of motion of these three particles become

$$m\ddot{x}_1 = -\frac{Tx_1}{a} + \frac{T(x_2 - x_1)}{a} = -\frac{2Tx_1}{a} + \frac{Tx_2}{a}$$

$$m\ddot{x}_2 = -\frac{T(x_2 - x_1)}{a} - \frac{T(x_2 - x_3)}{a} = \frac{Tx_1}{a} - \frac{2Tx_2}{a} + \frac{Tx_3}{a}$$

$$m\ddot{x}_3 = \frac{T(x_2 - x_3)}{a} - \frac{Tx_1}{a} = \frac{Tx_2}{a} - \frac{2Tx_3}{a}$$

This set of equations, written in matrix form, becomes

$$\begin{pmatrix} \ddot{x}_1 \\ \ddot{x}_2 \\ \ddot{x}_3 \end{pmatrix} = \frac{T}{ma} \begin{pmatrix} -2 & 1 & 0 \\ 1 & -2 & 1 \\ 0 & 1 & -2 \end{pmatrix} \begin{pmatrix} x_1 \\ x_2 \\ x_3 \end{pmatrix} \tag{6.25}$$

thus, incidentally, giving an example of a banded matrix being generated from a physical problem. The eigenvalues of the matrix are given by

$$\begin{vmatrix} -2-\lambda & 1 & 0 \\ 1 & -2-\lambda & 1 \\ 0 & 1 & -2-\lambda \end{vmatrix} = 0$$

the values being $\lambda = -2$, $-2 - \sqrt{2}$ and $-2 + \sqrt{2}$. The corresponding eigenvectors are given by

$$\begin{pmatrix} 1 \\ 0 \\ -1 \end{pmatrix}, \begin{pmatrix} 1 \\ -\sqrt{2} \\ 1 \end{pmatrix} \text{ and } \begin{pmatrix} 1 \\ \sqrt{2} \\ 1 \end{pmatrix}$$

in their simplest form. Hence, letting

$$\begin{pmatrix} x_1 \\ x_2 \\ x_3 \end{pmatrix} = \begin{pmatrix} 1 & 1 & 1 \\ 0 & -\sqrt{2} & \sqrt{2} \\ -1 & 1 & 1 \end{pmatrix} \begin{pmatrix} y_1 \\ y_2 \\ y_3 \end{pmatrix} \tag{6.26}$$

the equation of motion becomes

$$\begin{pmatrix} 1 & 1 & 1 \\ 0 & -\sqrt{2} & \sqrt{2} \\ -1 & 1 & 1 \end{pmatrix} \begin{pmatrix} \ddot{y}_1 \\ \ddot{y}_2 \\ \ddot{y}_3 \end{pmatrix} = \frac{T}{ma} \begin{pmatrix} -2 & 1 & 0 \\ 1 & -2 & 1 \\ 0 & 1 & -2 \end{pmatrix} \begin{pmatrix} 1 & 1 & 1 \\ 0 & -\sqrt{2} & -\sqrt{2} \\ -1 & 1 & 1 \end{pmatrix} \begin{pmatrix} y_1 \\ y_2 \\ y_3 \end{pmatrix}$$

so that
$$\begin{pmatrix} \ddot{y}_1 \\ \ddot{y}_2 \\ \ddot{y}_3 \end{pmatrix} = \frac{T}{ma} \begin{pmatrix} -2 & 0 & 0 \\ 0 & -2\sqrt{2} & 0 \\ 0 & 0 & -2+\sqrt{2} \end{pmatrix} \begin{pmatrix} y_1 \\ y_2 \\ y_3 \end{pmatrix} \qquad (6.27)$$

Letting $+\frac{2T}{ma} = \omega_1^2$, $\frac{T}{ma}(2-\sqrt{2}) = \omega_2^2$ and $\frac{T}{ma}(2+\sqrt{2}) = \omega_3^2$, where ω_1, ω_2 and ω_3 and all real, equation (6.27) can be rewritten as the three equations

$$\ddot{y}_1 = -\omega_1^2 y_1, \quad \ddot{y}_2 = -\omega_2^2 y_2 \text{ and } \ddot{y}_3 = -\omega_3^2 y_3$$

each being the simple harmonic equation for a different variable and so the solutions are

$$y_1 = b_1 e^{i\omega_1 t}, \quad y_2 = b_2 e^{i\omega_2 t}, \quad y_3 = b_3 e^{i\omega_3 t} \qquad (6.28)$$

The solution to the original problem in the chosen coordinates is then given by

$$\begin{pmatrix} x_1 \\ x_2 \\ x_3 \end{pmatrix} = \begin{pmatrix} 1 & 1 & 1 \\ 0 & -\sqrt{2} & \sqrt{2} \\ -1 & 1 & 1 \end{pmatrix} \begin{pmatrix} b_1 e^{i\omega_1 t} \\ b_2 e^{i\omega_2 t} \\ b_3 e^{i\omega_3 t} \end{pmatrix}$$

$$= b_1 \begin{pmatrix} 1 \\ 0 \\ 1 \end{pmatrix} e^{i\omega_1 t} + b_2 \begin{pmatrix} 1 \\ -2 \\ 1 \end{pmatrix} e^{i\omega_2 t} + b_3 \begin{pmatrix} 1 \\ 2 \\ 1 \end{pmatrix} e^{i\omega_3 t} \qquad (6.29)$$

Note that we can physically interpret equation (2.8) in the following way. The inverse of the matrix of eigenvectors is

$$\frac{1}{4} \begin{pmatrix} 2 & 0 & -2 \\ 1 & -\sqrt{2} & 1 \\ 1 & \sqrt{2} & 1 \end{pmatrix}$$

so that, from equation (6.26),

$$\begin{pmatrix} y_1 \\ y_2 \\ y_3 \end{pmatrix} = \frac{1}{4}\begin{pmatrix} 2 & 0 & -2 \\ 1 & -\sqrt{2} & 1 \\ 1 & \sqrt{2} & 1 \end{pmatrix}\begin{pmatrix} x_1 \\ x_2 \\ x_3 \end{pmatrix}$$

giving

$$y_1 = \frac{1}{2}(x_1 - x_3)$$

$$y_2 = \frac{1}{4}(x_1 - \sqrt{2}x_2 + x_3)$$

and

$$y_3 = \frac{1}{4}(x_1 + \sqrt{2}x_2 + x_3)$$

But y_1, y_2 and y_3 perform simple harmonic motion and so the relative values of x_1, x_2 and x_3 appearing in each of them must remain constant. This gives three possible types of oscillations and these are shown in diagrams 6.3, 6.4 and 6.5.

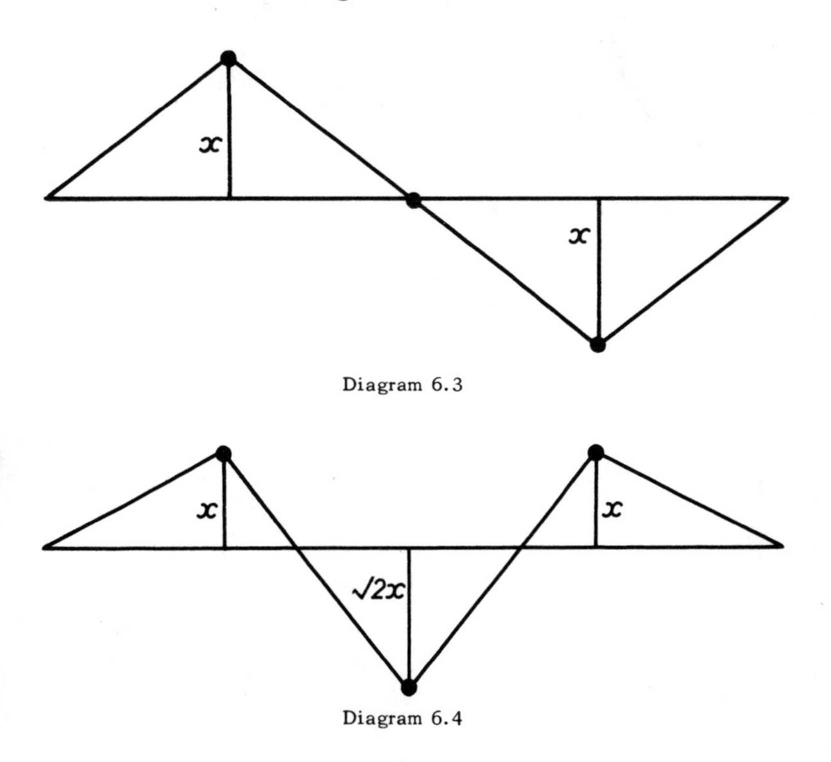

Diagram 6.3

Diagram 6.4

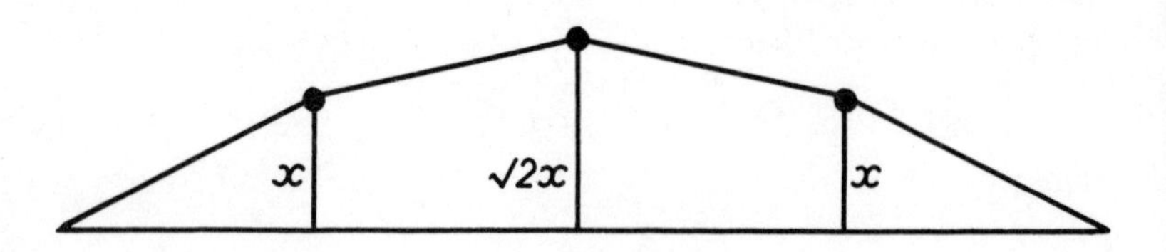

Diagram 6.5

The general solution is given by any linear combination of these NORMAL MODES OF OSCILLATION and this is what equation (6.29) gives. We see the close relationship between the natural frequencies (related to ω) and the eigenvalue

$$\left(= \frac{ma}{T}\,\omega^2\right)$$

This illustrates why it may be sufficient only to find smallest eigenvalues as this may determine the behaviour of the oscillation under certain circumstances. Alternatively, if the original equations were of the form

$$\mathbf{A}\dot{\mathbf{X}} = \mathbf{X}$$

then the largest eigenvalue would be important.

(b) In diagram 6.6 a typical electric circuit is illustrated.

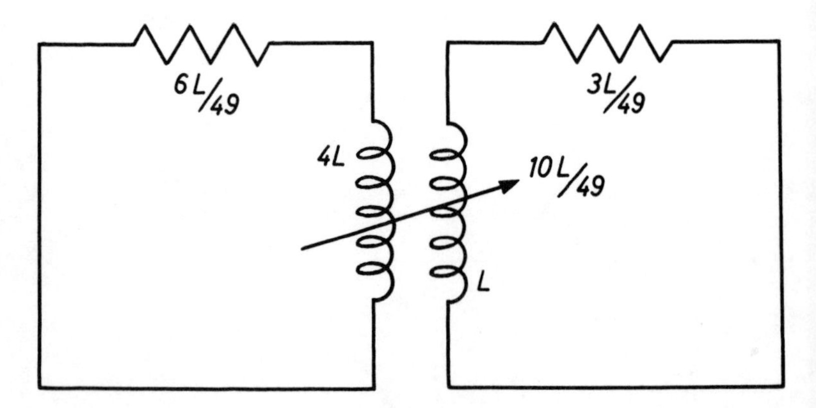

Diagram 6.6

One circuit contains an induction coil of self inductance L and a resistance of value $3L/49$ while the other coil contains a coil of self inductance $4L$ and a resistance of value $6L/49$. There is a mutual inductance of $10L/49$ between the coils. The circuit equations defining the currents I_1 and I_2 become

$$\frac{3L}{49} I_1 + \frac{10L}{49}\frac{dI_2}{dt} + L\frac{dI_1}{dt} = 0$$

and

$$\frac{6L}{49} I_2 + \frac{10L}{49}\frac{dI_1}{dt} + 4L\frac{dI_2}{dt} = 0 \tag{6.30}$$

We note that this method could also be used to find the complementary function part of the solution to the problem when sources of electromotive force are included, since the equations for the complementary function would be identical to equations (6.30) above.

Equations (6.30), written in matrix form, become

$$\begin{pmatrix} L & 10L/49 \\ 10L/49 & 4L \end{pmatrix}\begin{pmatrix} \dot{I}_1 \\ \dot{I}_2 \end{pmatrix} = \begin{pmatrix} -3L/49 & 0 \\ 0 & -6L/49 \end{pmatrix}\begin{pmatrix} I_1 \\ I_2 \end{pmatrix} \tag{6.31}$$

The matrix $\begin{pmatrix} -3L/49 & 0 \\ 0 & -6L/49 \end{pmatrix}$ clearly has an inverse and so the general theory for solving differential equations is applicable. We do not calculate this inverse but rather evaluate λ satisfying

$$\begin{vmatrix} L - \lambda(-3L/49) & 10L/49 \\ 10L/49 & 4L - \lambda(-6L/49) \end{vmatrix} = 0$$

or

$$\begin{vmatrix} 49 + 3\lambda & 10 \\ 10 & 196 + 6\lambda \end{vmatrix} = 0$$

This gives rise to the quadratic

$$18\lambda^2 + 882\lambda + 9504 = 0$$

which factorises to

$$(\lambda + 16)(\lambda + 33) = 0$$

When $\lambda = -16$, we have $\begin{pmatrix} 10 \\ -1 \end{pmatrix}$ as an eigenvector and when $\lambda = -33$, $\begin{pmatrix} 1 \\ 5 \end{pmatrix}$ is an eigenvector. Hence, choosing $\mathbf{Q} = \begin{pmatrix} 10 & 1 \\ -1 & 5 \end{pmatrix}$

and new variables $\begin{pmatrix} J_1 \\ J_2 \end{pmatrix}$ defined by

$$\begin{pmatrix} I_1 \\ I_2 \end{pmatrix} = \begin{pmatrix} 10 & 1 \\ -1 & 5 \end{pmatrix} \begin{pmatrix} J_1 \\ J_2 \end{pmatrix}$$

equation (6.31) becomes

$$L \begin{pmatrix} 1 & 10/49 \\ 10/49 & 4 \end{pmatrix} \begin{pmatrix} 10 & 1 \\ -1 & 5 \end{pmatrix} \begin{pmatrix} \dot{J}_1 \\ \dot{J}_2 \end{pmatrix} = -L/49 \begin{pmatrix} 3 & 0 \\ 0 & 6 \end{pmatrix} \begin{pmatrix} 10 & 1 \\ -1 & 5 \end{pmatrix} \begin{pmatrix} J_1 \\ J_2 \end{pmatrix}$$

Multiplying both sides of this equation by $\mathbf{Q}^{-1} \begin{pmatrix} 3 & 0 \\ 0 & 6 \end{pmatrix}^{-1}$ yields

$$\begin{pmatrix} -16 & 0 \\ 0 & -33 \end{pmatrix} \begin{pmatrix} \dot{J}_1 \\ \dot{J}_2 \end{pmatrix} = \begin{pmatrix} 1 & 0 \\ 0 & 1 \end{pmatrix} \begin{pmatrix} J_1 \\ J_2 \end{pmatrix}$$

This equation gives

$$\dot{J}_1 = -1/16\, J_1 \text{ or } J_1 = ae^{-t/16}$$

and

$$\dot{J}_2 = -1/33\, J_2 \text{ or } J_2 = be^{-t/33}$$

so that the solution to the equations in terms of currents is given by

$$\begin{pmatrix} I_1 \\ I_1 \end{pmatrix} = \begin{pmatrix} 10 & 1 \\ -1 & 5 \end{pmatrix} \begin{pmatrix} ae^{-t/16} \\ be^{-t/33} \end{pmatrix}$$

or

$$\begin{pmatrix} I_1 \\ I_2 \end{pmatrix} = a \begin{pmatrix} 10 \\ -1 \end{pmatrix} e^{-t/16} + b \begin{pmatrix} 1 \\ 5 \end{pmatrix} e^{-t/33}$$

If there were many circuits present the same method would apply; the important eigenvalues would then be the largest ones if positive or the smallest (in modulus) if negative.

One differential equation of order n

The above method of solving a differential equation can be modified to solve an equation of the form

$$\frac{d^n y_1}{dx^n} + a_n \frac{d^{n-1} y_1}{dx^{n-1}} + \dots a_2 \frac{dy_1}{dx} + a_1 y_1 = 0$$

Let

$$\frac{dy_1}{dx} = y_2$$

$$\frac{dy_2}{dx} = y_3$$

$$\dots = \dots$$

$$\frac{dy_{n-1}}{dx} = y_n$$

The original equation then becomes

$$\frac{dy_n}{dx} = -a_n\, y_n - a_{n-1}\, y_{n-1} \dots - a_1 y_1$$

and so the whole equation is equivalent to solving

$$\begin{pmatrix} \dot{y}_1 \\ \dot{y}_2 \\ \dots \\ \dot{y}_n \end{pmatrix} = \begin{pmatrix} 0 & 1 & 0 & \dots & 0 \\ 0 & 0 & 1 & \dots & 0 \\ \dots & \dots & \dots & \dots & \dots \\ -a_1 & -a_2 & -a_3 & \dots & -a_n \end{pmatrix} \begin{pmatrix} y_1 \\ y_2 \\ \dots \\ y_n \end{pmatrix}$$

which is now identical to the set of equations given by (6.13) and can be solved in a similar manner to them.

Deformation of a block of material

Consider a deformation of a block of material such that any point with coordinates (x, y, z) is changed to the point with coordinates (x', y', z') and the relationship between the two sets of coordinates is given by

$$\begin{pmatrix} x \\ y \\ z \end{pmatrix} = \begin{pmatrix} a_{11} & a_{12} & a_{13} \\ a_{21} & a_{22} & a_{23} \\ a_{31} & a_{32} & a_{33} \end{pmatrix} \begin{pmatrix} x' \\ y' \\ z' \end{pmatrix}$$

or

$$\mathbf{X} = \mathbf{A}\,\mathbf{X}' \qquad (6.32)$$

Suppose further that the deformation is such that will make $\mathbf{A}$ a symmetric matrix as will be the case for many of the deformations that occur in practice.

As $\mathbf{A}$ is a symmetric matrix, its eigenvalues and eigenvectors can be calculated and the matrix of eigenvectors, which is orthogonal, will transform $\mathbf{A}$ to the diagonal matrix whose elements are the eigenvalues, Λ, in the following way

$$\mathbf{Q}^T \mathbf{A} \mathbf{Q} = \Lambda$$

Now an orthogonal matrix has been shown to be a matrix that represents the rotation of a system of axis. Let us therefore rotate the axis of reference of the deformed material so that

$$\mathbf{X}' = \mathbf{Q} \mathbf{X}'' \tag{6.33}$$

The relationship between the original coordinate system and this new rotated set of axis then becomes, on combining equations (6.32) and (6.33),

$$\mathbf{X} = \mathbf{A} \mathbf{Q} \mathbf{X}''$$

so that

$$\mathbf{Q}^T\mathbf{X} = \mathbf{Q}^T \mathbf{A} \mathbf{Q} \mathbf{X}'' = \Lambda \mathbf{X}'' \tag{6.34}$$

Also, from equation (6.32) on transposing

$$\mathbf{X}^T = \mathbf{X}^T \mathbf{A}^T$$

or, as $\mathbf{A}$ is a symmetric matrix,

$$\mathbf{X}^T = \mathbf{X}'^T \mathbf{A}$$

Equation (6.33) on transposing, gives

$$\mathbf{X}'^T = \mathbf{X}''^T \mathbf{Q}^T$$

and so

$$\mathbf{X}^T = \mathbf{X}''^T \mathbf{Q}^T \mathbf{A}$$

which gives

$$\mathbf{X}^T\mathbf{Q} = \mathbf{X}'^T \mathbf{Q}^T \mathbf{A} \mathbf{Q} = \mathbf{X}''^T \Lambda \tag{6.35}$$

Combining equations (6.34) and (6.35) gives an interesting interpretation of the meaning of eigenvalues in this context. It gives

$$\mathbf{X}^T\mathbf{Q} \mathbf{Q}^T\mathbf{X} = \mathbf{X}''^T \Lambda^2 \mathbf{X}''$$

or, since $\mathbf{Q}$ is an orthogonal matrix,

$$\mathbf{X}^T\mathbf{X} = \mathbf{X}''^T \Lambda^2 \mathbf{X}'' \tag{6.36}$$

Consider now a spherical cavity of radius a and centre the origin in the block of material. Its equation is therefore

$$\mathbf{X}^T\mathbf{X} = a^2$$

In the deformed material this cavity will have the equation

$$\mathbf{X}''^T \Lambda^2 \mathbf{X}'' = a^2 \tag{6.37}$$

on using equation (6.36).

Writing this equation out in full gives

$$(x''\ y''\ z'') \begin{pmatrix} \lambda_1^2 & 0 & 0 \\ 0 & \lambda_2^2 & 0 \\ 0 & 0 & \lambda_3^2 \end{pmatrix} \begin{pmatrix} x'' \\ y'' \\ z'' \end{pmatrix} = a^2$$

and, on carrying out the necessary multiplication, it becomes

$$\lambda_1^2 x''^2 + \lambda_2^2 y''^2 + \lambda_3^2 z''^2 = a^2$$

or

$$\frac{x''^2}{\left(\frac{a}{\lambda_1}\right)^2} + \frac{y''^2}{\left(\frac{a}{\lambda_2}\right)^2} + \frac{z''^2}{\left(\frac{a}{\lambda_3}\right)^2} = 1$$

which is, of course, the equation of an ellipsoid with semi-axes a/λ_1, a/λ_2 and a/λ_3. The three eigenvalues therefore determine the amount of magnification that occurs in the deformation from the spherical stage to the ellipsoidal shape. Further, the matrix of eigenvectors, **Q**, determines the directions which the axes of the ellipsoid, and therefore the directions of the principal stresses, make with the original directions of the axes.

FURTHER READING

A more advanced treatment of matrix theory and its extensions can be found in the following books:

Daniel T. Finkbeiner, *Introduction to Matrices and Linear Transformations*, 1966.

A.M. Tropper, *Linear Algebra*, 1968.

Seymour Lipschutz, *Theory and Problems of Linear Algebra*, 1968

A large collection of exercises and worked examples at roughly the level of this text can be found in:

Frank Ayeres, *Theory and Problems of Matrices*, 1962

The numerical aspects of the work that has been discussed, together with more sophisticated methods, may be found in:

I. Khabaza, *Numerical Analysis*, 1965
S. D. Conte, *Elementary Numerical Analysis*, 1965
P. A. Stark, *Introduction to Numerical Methods*, 1970
A. Ralston, *A First Course in Numerical Analysis*, 1965

The application of matrix theory to specific problems in science and engineering are discussed in:

A. M. Tropper, *Matrix Theory for Electrical Engineering Students*, 1962
W. G. Bickley and R. S. H. G. Thompson, *Matrices, Their Meaning and Manipulation*, 1964

INDEX

Addition of matrices, 20
Adjoint matrix, 40
Adjugate matrix, *see* Adjoint
Anti-hermitian matrix, 26
Anti-symmetric matrix, 25
Applications, 121
Approximate inverse, 48
Arbitrary vector, for eigenvalues, 102
Axis, rotation of, 93

Back substitution, 65
Banded matrix, 22
Bending strutt, 125

Characteristic equation, 78
Characteristic root, *see* Eigenvalue
Cofactors, 39
Column: eigenvector, 78; of a matrix, 13; vector, 18

Data analysis, 123
Decomposition, 67
Definition of matrices, 13
Deformation of materials, 124
Determinant of a matrix, 17; multiplication, 17
Diagonal-leading, 21
Diagonal matrix: definition, 21; reduction to, 81; orthogonal reduction to, 86
Differential equations, 130

Eigenvalues: definition, 78; determining largest, 102; determining second, 108; eliminating an, 111; geometric interpretation, 92; Jacobi method, 116; product of, 78; Rayleigh–Schwartz method, 110; smallest, 115; sum of, 78
Eigenvector: column, 78; definition, 78; matrix of, 81; normalized; 87; Rayleigh–Schwartz, 110; row, 78

Elements of a matrix, 13
Equality of matrices, 14
Exponential of a matrix, 98

Function of a matrix, 97

Gaussian elimination, 63
Gauss–Jordan determination of the inverse, 44
Gauss–Jordan comparison with other methods, 44
Gauss–Seidel iterative method, 72
Geometric interpretation of eigenvalues, 92

Hermitian matrix, 25
Homogeneous equations: definition, 57; solution, 61

Inverse: definition, 22; determination, 40; existence of, 40; Gauss–Jordan, 44; improving approximation, 48; of product, 43; of transpose, 43; partitioning, 52
Iterative method: Gauss–Seidel, 72; simple, 70

Jacobi method for eigenvalues, 116
Jordan, *see* Gauss–Jordan

Latent root, *see* Eigenvalue
Leading diagonal, 21
Linear dependence: of eigenvectors, 81; of vectors, 59
Lower triangular matrix: definition, 24; in a product, 30, 34

Matrix: addition, 20; adjoint, 40; adjugate, 40; anti-hermitian, 26; anti-symmetric, 25; banded, 22; definition, 13; determinant of, 17; diagonal, 21; elements of, 13; exponential, 98; function of, 97; hermitian, 25; inverse, 22;

lower triangular, 24; minor, 39; multiplication of, 14; non-singular, 17; null, 21; order of, 13; orthogonal, 27; of eigenvectors, 81; product of, 14; partitioned, 50; rank of, 61; reciprocal, 22; singular, 17; skew hermitian, *see* Anti-hermitian; skew symmetric, *see* Anti-symmetric; spur, 78; square, 13; stiffness, 128; sub, 50; sum of, 20; symmetric, 24; trace of, 78; transpose of, 18; unit, 22; unit lower triangular, 24; unit upper triangular, 23; upper triangular, 23; zero, *see* Null

Minors, 39

Multiplication: by a scalar, 15; of determinant, 17; of matrices, 14

Non-homogeneous: definition, 57; solution, 62

Non-singular matrix, 17

Non-trivial solutions, 57

Norm of a vector, 27

Normal modes of oscillation, 136

Null matrix, 21

Order of a matrix, 13

Orthogonal: matrix, 27; reduction; 86

Orthogonality of vectors, 27

Partitioned matrix, 50

Partitioning for the inverse, 52

Pivoting, 46

Pole, *see* Eigenvector

Product: of determinants, 17; of eigenvalues, 78; of inverse, 43; of matrices, 14; of reciprocal, 43; of transpose, 18

Quadric, 93

Rank; homogeneous system, 61; non-homogeneous system, 62

Rayleigh–Schwartz method, 110

Reciprocal: calculation, 40; calculation by Gauss–Jordan, 44; existence of, 40; of a matrix, 22; product of, 43; of transpose, 43

Reduction to a diagonal matrix, 61

Rotation of axis, 93

Row: eigenvector, 78; of a matrix, 13; vector, 18

Scalar: multiplication, 15; product of a vector, 12

Singular matrix, 17

Skew hermitian matrix, *see* Anti-hermitian matrix

Skew symmetric matrix, *see* Anti-symmetric matrix

Spectral resolution, 96

Spur, 78

Square matrix, 13

Stiffness matrix, 128

Structure displacement, 128

Submatrices, 50

Sum of: eigenvalues, 78; matrices, 20

Symmetric matrix, 24

Trace, 78

Transpose: of the inverse, 43; of the matrix, 18; of the product, 18; of the reciprocal, 43

Trivial solutions, 57

Unit: lower triangular matrix, 24; matrix, 22; upper triangular matrix, 23

Upper triangular matrix, definition, 23

Vector, column, 18; norm of, 27; orthogonal, 27; row, 18; scalar product, 12

Wheatstone bridge circuit, 126

Zero matrix, *see* Null matrix